NON-LINEAR DYNAMICS IN GEOPHYSICS

WILEY-PRAXIS SERIES IN GEOPHYSICS
Series Editor: **Philippe Blondel, M.Sc., Ph.D.**
Southampton Oceanography Centre, Southampton, UK

Knowledge of the physics of our Earth has steadily advanced since the days of Cuvier and Darwin, and a huge step forward took place with the theory of plate tectonics in the 1970s. Further significant progress has resulted from discoveries made throughout the Solar System during the past two decades. However, many important and fundamental questions remain unanswered. Is our planet unique? How does it function? How can we improve our understanding of its mechanisms?

The Wiley-Praxis Series in Geophysics has been developed to make accessible state-of-the-art knowledge about the Earth and geophysics, and the investigative and analytical techniques employed, and to present this information in a logical and cohesive way. In addition to texts of a more general nature, titles in the series will also cover topical themes, dedicated to one particular subject or field of research. Geophysics is a young science, and new areas of study arise almost every day.

The target audience for the books in the series is large, international and with varying levels of scientific and technical knowledge. The more general books will appeal to undergraduate students and academics, but will also be of interest to specialists wishing to keep abreast of new results and theories. More specialised works, for example on geophysical techniques, will appeal particularly to professional scientists, and also postgraduate students and researchers wishing to refine their expertise.

For further details of the book listed below and ordering information, why not visit the Praxis Web Site at http://www.praxis-publishing.co.uk

NON-LINEAR DYNAMICS IN GEOPHYSICS
Jacques Dubois, Professor of Geophysics, Paris University, France

NON-LINEAR DYNAMICS IN GEOPHYSICS

Jacques Dubois
Professor of Geophysics, Paris University, France

JOHN WILEY & SONS
Chichester • New York • Weinheim • Brisbane • Singapore • Toronto

Published in association with
PRAXIS PUBLISHING
Chichester

Copyright © 1998 Praxis Publishing Ltd
The White House,
Eastergate, Chichester,
West Sussex, PO20 6UR, England

Original French edition, *La dynamique non linéaire en physique du globe*,
published by MASSON, Editeur, Paris, © 1995.

English language edition published in 1998 by
John Wiley & Sons Ltd
in association with Praxis Publishing Ltd.
This work has been published with the help of the French Ministère de la Culture.

All rights reserved

No part of this book may be reproduced by any means,
or transmitted, or translated into a machine language
without the written permission of the publisher

Wiley Editorial Offices

John Wiley & Sons Ltd, Baffins Lane,
Chichester, West Sussex, PO19 1UD, England

John Wiley & Sons, Inc., 605 Third Avenue,
New York, NY 10158-0012, USA

Wiley-VCH Verlag GmbH, Pappelallee 3,
D-69469 Weinheim, Germany

Jacaranda Wiley Ltd, G.P.O. 33 Park Road, Milton,
Queensland 4001, Australia

John Wiley & Sons (Asia) Pte Ltd, 2 Clementi Loop #02-01,
Jin Xing Distripark, Singapore 12981

John Wiley & Sons (Canada) Ltd, 22 Worcester Road,
Rexdale, Ontario, M9W 1L1, Canada

Library of Congress Cataloguing-in-Publication Data

A catalogue record for this book is available from the British Library

ISBN 0-471-97853-1

Translated by Dr Philippe Blondel, Senior Scientist, Southampton Oceanography Centre,
Southampton

Printed and Bound in Great Britain by MPG Books Ltd, Bodmin

Table of Contents

INTRODUCTION	**1**
1 Natural Phenomena and Non-Linear Dynamics	**1**
2 Outline of This Work	**3**
1. FRACTAL AND MULTIFRACTAL THEORY	**9**
1.1 Notion of Measure	**9**
1.1.1 Measuring by Counting	10
1.1.2 Mass of a Point	10
1.1.3 Lebesgue Measure in $\mathcal{R}^1$ and $\mathcal{R}^n$	10
1.1.4 Hausdorff Measure	11
1.2 Notion of Dimension	**13**
1.2.1 Hausdorff Dimension	13
1.2.2 Another Definition of the Hausdorff Dimension	15
1.3 Fractal Sets	**17**
1.3.1 Definitions	17
1.3.2 Triadic Cantor Set	18
1.3.3 Dimension of the Triadic Cantor Set	20
1.3.4 Some Characteristic Properties of the Triadic Cantor Set	21
1.3.5 Random Cantor Set	21
1.3.6 Other Examples of Fractal Sets	23
1.3.7 Computation of the Fractal Dimension	25
1.3.8 A Few Properties of Fractal Sets: Projection, Product, Intersection	27
1.4 Dynamic Systems	**31**
1.4.1 The Harmonic Oscillator	31
1.4.2 The Simple Pendulum	32
1.4.3 Coupled Harmonic Oscillators	33

1.4.4 Systems with Central and Non-Central Force Fields	34
1.4.5 Summary	37
1.5 Limit Cycles and Attractors	**37**
1.5.1 The Attraction Process	37
1.5.2 Area Contraction	38
1.5.3 Aperiodic Attractors, Chaotic Regime, Strange Attractors	39
1.5.4 Practical Search for an Attractor and its Dimension	41
1.5.5 Examples of Strange Attractors	44
1.5.6 Lyapunov Exponents	49
1.5.7 Poincaré Sections and First-Return Maps	53
1.5.8 Logistic Map, Sub-Harmonic Cascade and Chaos	55
1.6 Multifractals and Wavelet Transforms	**59**
1.6.1 Notion of Generalised Dimension	59
1.6.2 Multifractals - Theory and Practice	62
1.6.3 Wavelet Transform	67
1.7 Notion of Chaos	**71**
1.7.1 The Path to Chaos - Hopf Bifurcation	71
1.7.2 From the T^2 toroid to the T^3 Toroid - Theory of Ruelle-Takens	76
1.7.3 The Curry and Yorke Model	77
1.7.4 Definitions of Chaos	80
2. APPLICATIONS TO GEOPHYSICS	**81**
2.1 Geomorphology	**82**
2.1.1 Observations	82
2.1.2 Fractal Self-Affine Topographic Profiles	83
2.1.3 2-D Representation	85
2.2 Fragmentation, Fracturation, Tectonics, Percolation	**87**
2.2.1 Fragmentation	87
2.2.2 Rock Fracturation	95
2.2.3 Tectonics - Study of Surface Faults	101
2.2.4 Percolation	103
2.3 Seismicity - Gutenberg-Richter Law	**107**
2.3.1 Power Law or Poisson Law ?	107
2.4 Fragmentation - Tectonics - Seismicity	**113**
2.5 Probabilistic Approach - Applications	**116**
2.5.1 Observables	116
2.5.2 Application to Volcanic Eruptions	118
2.5.3 Cantor Dust	119
2.5.4 Correlation Function	120
2.5.5 First-Return Map	122
2.5.6 Multifractal Analyses	124
2.6 Self-Organised Criticality: SOC	**126**
2.7 The Russian Method: Algorithms M8 and CN	**129**
2.7.1 Principles	130
2.7.2 The M8 Algorithm	130
2.7.3 The CN Algorithm	133
2.8 Geomagnetism	**137**
2.8.1 Inversions of the Earth's Magnetic Field	137
2.8.2 Temporal Variations of the Magnetic Field Vector	143

3. PERSPECTIVES — 149
3.1 General Points — 149
3.2 Chaotic Time Series — 151
3.3 Fractals, Multifractals and Wavelet Transforms — 154
3.4 Self-Organised Criticality (SOC) — 154
3.5 The Problem of Short Series — 156
3.6 Mastering and Controlling Chaos — 157

4. ILLUSTRATIONS OF NON-LINEAR DYNAMICS — 159
4.1 Fractal Sets — 159
4.1.1 The Olbers Paradox — 160
4.1.2 Iterated Function Systems (IFS) — 160
4.1.3 Julia Set - Mandelbrot Set — 165
4.2 Dynamic Systems — 169
4.2.1 Sinaï's Billiards — 169
4.2.2 The Baker Transform — 169
4.3 Self-Organised Criticality — 171
4.3.1 The Game of Life — 171

ANNEX A - FRACTAL THEORY — 173
A.1 Measure Theory — 173
A.1.1 Definition of the Hausdorff Measure — 173
A.1.2 About the Notions of Measure and Dimension — 174
A.1.3 About the Hausdorff Dimension — 175
A.2 Fractals — 175
A.2.1 Hölder Function, Hölder Condition, Lipschitz Map — 175
A.2.2 Mandelbrot Set on Quaternions — 177
A.2.3 Fractal Projections — 178
A.3 Dynamic Systems and Chaos — 181
A.3.1 Degree of Freedom, Phase Space, Phase Trajectories — 182
A.3.2 Chaos Applied to Dynamic Systems — 182
A.3.3 Control of Chaotic Systems — 183
A.4 Multifractals — 188
A.4.1 The Histogram Method — 188
A.4.2 The Moments Method — 189

ANNEX B - APPLICATIONS TO GEOPHYSICS — 195
B.1 Geomorphology — 195
B.1.1 Fractal Analysis of River Networks — 195
B.1.2 Scale Invariance and Geoid Roughness — 198
B.2 Fragmentation, Fracturation, Tectonics and Percolation — 203
B.2.1 Fracturation and an Anisotropic Model — 203
B.2.2 Percolation, a Geometric Disorder — 210
B.3 Seismology - Russian Methods — 219

B.3.1 Prediction Method of P. Shebalin and I. Rotwain 219
B.4 Geomagnetism 224
B.4.1 Study of a Long Magnetic Series 224
B.5 Palaeontological Studies of Evolution 232

Bibliography **235**
Index **253**

The colour plate section is between pages 192 and 193

Foreword to the French Edition

The lectures of Professor J. Dubois present a long overdue move - difficult, bold and successful - to narrow the current gap in understanding of critical phenomena in solid Earth and of its instabilities in general. This gap, crucial both for scientific and practical reasons - for perception and for survival - separates the dominating trends in solid Earth sciences from the recent recognition of deterministic chaos in the dynamics of lithosphere.

This recognition emerged six or seven years ago. Its overwhelming potential consequences can be best appreciated in historical perspective. Two great Frenchmen, Pierre-Simon de Laplace in the last third of the XVIIIth century and Jules Henri Poincaré at the turn of the XXth century, developed the apparently opposite though in fact complementary attitudes to the dynamics of Nature, "to the flow of change". Poincaré summarized both attitudes as follow:

- *"If we knew exactly the laws of Nature and the situation of the Universe at the initial moment, we could predict exactly the situation of that same Universe at a succeeding moment"* (and to reconstruct a preceding moment too).

That is what Laplace believed in, and for good Newtonian reasons. But, points out Poincaré a century and a quarter later:

- " *... it is not always so; it may happen that small differences in the initial conditions produce very great ones in the final phenomena. A small error in the former will produce an enormous error in the latter. Prediction becomes impossible, and we have the fortuitous phenomena*".

Such phenomena arise in nature and society when the cause-and-effect relations are non-linear, even though deterministic. This combination generates erratic complicated time-space structures that fascinate us, for example in the roaring flame or turbulent waterfall. Naturally such behaviour is called deterministic chaos; it is studied by a scientific discipline called non-linear dynamics .

A simple example of deterministic chaos is the movement of balls in the game of billiards. The smallest change in the direction of the initial shot will change the whole subsequent motion of all the balls, and the difference will not attenuate but grow in time, amplified by each collision (non-linearity in this case is due to the curvature of the ball's surface). Therefore, though the motion of the balls is completely determined by the Newton laws, its precise prediction is impossible, since it would require absurd precision of initial conditions.

"Though this be madness, there is method in't" (Shakespeare): chaotic behavior has its own intrinsic regularities, reviewed in the first part of this book: bifurcations, typical spatio-temporal behaviour patterns; their alternation in time ("intermittence"); concentration of a process on some configurations in the phase space ("strange attractors"); transition of the system to a higher level of inequilibrium ("dissipative structures"); collective behaviour or self-organisation eluding explanation on elementary level (as "communication between molecules", an expression of Prigogine).

The last two features make chaotic systems prone to critical phenomena or phase transitions, that is, abrupt overall changes of the whole system. Development of such changes often follows certain scenarios, common for rather diverse systems.

The discoveries of Poincaré warn us that it is not possible to understand a chaotic system by breaking it apart; that the quest for more and more details is futile for prediction. As soon as this impossibility is recognised, chaotic processes may become open for understanding, for prediction, and even for control - in their own sense, with their own limitations and by their own laws.

Two approaches to prediction deserve potential attention:

- Recognising a typical behaviour pattern, we know - within limits - how it will evolve in the immediate future. For example, the straws in the wind may tell us that the storm is coming.

- Smoothing away (by averaging) the details of a chaotic process, we may achieve even long-term predictability, impossible prior to averaging. Consider, for example, the heart (which is also a chaotic system). The approach of a heart attack can be reliably diagnosed, within limits, by the strongly-averaged characteristics of the organism, such as ECG patterns, blood pressure, temperature. To start diagnosis by monitoring each of the cells participating in the function of the heart, would be obviously futile.

These examples may sound trivial, but in each newly recognised chaotic system it is not easy to find what are the scenarios of development, or which specific averaging yields predictability. Anyhow, chaotic systems have to be explored in analytical mode - starting from the whole and proceeding to consecutively divided parts.

For a long time, mainly, if not only, mathematicians were interested in deterministic chaos. This was changed in 1963, when Edward Lorenz discovered a strange attractor in a common natural process - thermal convection in the atmosphere, which shows an astonishing tendency to self-organise into a grid of convection cells.

Since then, deterministic chaos was recognized in exceedingly diverse systems: from very simple, like the non-linear pendulum or dripping faucet, to the brain, stock market, stars, galaxies, ocean and atmosphere, inner core of the Earth ... It looks like the universe of chaos is our universe proper, where linear systems are singular exceptions, like a telegraph pole in the tropical forest.

We can now turn to the Earth's lithosphere. How is it related to all, if anything, of the above?

The answer is simple, though somewhat obscured by the diversity of observations and the multitude of models: *the lithosphere presents a hierarchical non-linear dissipative system*. This is obvious, come to think of it.

- Why hierarchical? Since it is consecutively divided into a hierarchy of volumes ("blocks"), which move relatively to each other. The largest blocks are tectonic plates; they are divided into platforms, shields, mountain regions, etc; in 15-20 divisions, we come to the grains of rocks and may proceed even to the crystals. The blocks are separated by the relatively thin boundary layers: fault zones and faults at the top of the hierarchy, interfaces between the grains at its bottom. At least the fault zones feature similar hierarchy, with more frequent division.

- Why non-linear? Because of the strong instability in the interaction of blocks. More specifically, the relative movement of the blocks is controlled by a multitude of mechanisms, concentrated in boundary layers: migration of fluids, reducing the friction; stress corrosion, causing fatigue; petrochemical transitions; fracturing, buckling and plastic deformations, etc. All these mechanisms interact along and across the hierarchical structure of the lithosphere. No mechanism is always dominant, so that the rest can be ignored. Even a grain of rocks - the prime element of which the lithosphere is assembled - cannot be described by a simple model. It may act as a visco-elastic element, as an aggregate of crystals, as a source or absorber of fluids, and/or energy and/or volume, etc., playing several of these roles simultaneously.

To account, blow by blow, for each component of such complexity, would be unrealistically complicated; moreover, their interaction may create a new quality: a non-linear system is larger than the sum of its parts.

This complexity may be overlooked and simple models (e.g. visco-elastic ones) applied in certain specific cases: far from instability, in sufficiently slow time-scale, etc. However, it defeats at least my imagination how this complexity could be ignored for so long in the study of critical phenomena, be it geological disasters, super-large mineral deposits or self-destruction of mega-cities.

- Why dissipative? Because lithospheric structures which have reached sub-critical conditions tend to self-organise, generating such conditions in the structures of consecutively larger dimensions - again, up to a limit.

What did we really gain, evoking non-linear dynamics? What is its answer to the chaotic complexity of the lithosphere? We are so far in the most challenging initial stage - the heuristic search and modelling of basic regularities necessary to construct the fundamental equations, defining the lithosphere as a non-linear system.

Even in this infant stage, the concepts and the experience of non-linear dynamics led to significant reconsideration of existing data and yielded some new paradigms: the dynamics of different parts of the lithosphere is correlated over distances much larger than traditional models could explain, and intuition accept; after proper averaging, many processes, including the symptoms of approaching instability, became similar in diverse tectonic environments, and in this wide range of space and energy scales (for seismicity, this similarity extends from subduction zones to transform faults, to active platforms, to areas of induced activity); the structure of the lithosphere is fractal; many important traits of short-term dynamics of the lithosphere are reproduced in simple models of interacting elements, which do not include specifically geological features.

What may be even more important, non-linear dynamics highlighted the key problems, which until recently enjoyed little awareness: dimensionality of attractor; type of transition to chaos; relation between the time and space scales; limits of similarity and self-similarity; fractal dimensionality, etc.

The lithosphere seems to belong to the little-known type of non-linear systems with at least two distinctive traits: it has an intermediate number of degrees of freedom; it may remain near the critical state even after a large discharge of energy ("self-organised criticality").

In 20-20 hindsight, the non-linear dynamics approach to the lithosphere was well prepared by the following developments of the past 3-4 decades:

- Global geophysical networks were established, to a large extent, on defence budgets. For the study of Earth's dynamics, it enhanced the possibilities to proceed from the whole (the Earth) to its parts and to explore the limits of similarity.

- Methods of exploratory data analysis, especially pattern recognition, were developed, providing a powerful tool to identify the basic regularities in observed processes and to isolate illusionary ones, which is very easy to imagine in chaotic phenomena (e.g. in patterns of tea leaves).

- Plate tectonics provided a unifying framework for the dynamics of different parts of the lithosphere, on high levels of its hierarchy.

- A multitude of mechanisms of instability in the lithosphere was discovered in basic and applied research, making inevitable the non-linear dynamics approach.

Still the gap mentioned at the beginning of this foreword exhibits little tendency to disappear. The new trends discussed above are brought to Earth sciences mainly by physicists and mathematicians, who are discovering for themselves the solid Earth, while the bulk of the solid Earth sciences community remains (with notable exceptions) uninvolved. Massive accumulation of more and more detailed data is still a dominating trend in basic and applied research of the solid Earth: according to the French proverb "everybody is satisfied with his wisdom, but nobody is satisfied with his wealth" (counted in that case in observational systems, databases and computing facilities).

This is not unusual for the history of science. Even the discovery of Lorenz was accepted by his own community only after 15-20 years of half-benign neglect while triumphing in a score of other fields. At present the study of the human genome possibly also exhibits a sweeping quest for the details of a hierarchical chaotic system.

Probably contributing to inertia is the fact that the R&D in geological disasters is a cost-plus operation, not unreminiscent of the former Soviet economy, and with a fatalistic attitude to possible results the growing cost of data accumulation and processing may seem a symptom of progress.

No doubt, the fatalism will be soon replaced by intense, exaggerated expectations, due to the simple reason that understanding, prediction and control of critical phenomena in the lithosphere became crucially important for the very survival of mankind: survival against the combined threat of natural disasters, deterioration of ecology and depletion of natural resources.

Our vulnerability to these calamities is rapidly growing, so that their danger is considered now "*as great as any posed by Hitler, Stalin and the atom bomb*" (G. Wiesener). Huge spendings do not contain this growth. Deliverance obviously requires basic research, and non-linear dynamics is a natural part of it.

Neither this book nor this preface are concerned with criticism. Not at all - rather they consider large additional possibilities in R&D.

Another French proverb says: "Pour commencer il faut commencer" ("To begin, one must begin"), and this book is written in a highly constructive spirit: it outlines the evidence for the patterns of non-linear dynamics in solid Earth; it is addressed directly to the Earth science community, to an observationalist and data analyst; and in this community it goes first of all to its future, to the university students.

Volodya Keilis-Borok

Russian Academy of Science, Moscow

Acknowledgments

This book was first published with Masson (Paris), in the series of textbooks edited by Albert Tarantola for the M. Res. in Geophysics of the Institut de Physique du Globe de Paris (IPGP). It contains the lectures given in the option "Non-linear dynamics in geophysics" during the academic years 1992-1993 and 1993-1994.

The project goes back to autumn 1991, where at the prayer of A. Tarantola I had started writing some notes, finished in July 1992 and used by students the next year. Previously, since 1989, I had taught students the basics of fractal geometry, applied mainly to geomorphology and seismology.

My personal beginnings in non-linear dynamics go back to the beginning of the 1980s. More precisely, my first contact with the notion of fractal objects stems from a conversation with Professor Vladimir Pisarenko, of the Institute of Earth Physics in Moscow. Staying in Paris at CNRS, he gave me my first lesson about the non-integer dimension of Brittany's coastline. Later came the first publications, such as the one by Allègre, Le Moüel and Provost in 1982, which showed the interest of the fractal approach for the study of rock fracturation, and of course the reading of the classic books "Fractal Objects" by Mandelbrot, and "L'ordre dans le Chaos" ("Order in Chaos") by Bergé, Pomeau and Vidal. My first research in the domain started at this time with earthquakes series in the Tonga Islands, and the eruptive series of the Piton de la Fournaise. Several students trusted me and started Ph.D. theses on the matter. I want to hold up as an example Lassaad Nouaili, who must have been the first student in France to finish a Ph.D. (in 1989) which was entirely devoted to the fractal study of earthquakes in subduction zones. Jean-Pierre Kahane, mathematician and formerly president of the University of Paris-Sud Orsay, had accepted to be part of the committee, and had brought a definite help in the rigorous manipulation of fractal definitions and properties. Other Ph.D. theses followed, such as Anne Sauron-Sornette (1990), about scaling laws in fissured environments, Anicet Beauvais (1991) about the geomorphology of rivers in Central Africa, Béatrice Ledésert about rock fracturation. I would like to thank here these young scientists who understood the interest of these new approaches, despite the reticent advice of many older scientists. Among the newcomers, I am keen to mention Valérie Ballu, who studied the "roughness" of the seafloor as a result of determinist dynamic systems, and Lionel Hongre, who investigated the attractors of dynamic systems generating the variations of the Earth's magnetic field with the Lyapunov exponents, Pascal Seilhac who showed the multifractal properties of volcanic eruptions, David Aubert and Jean Battaglia who worked on the applications of non-linear techniques. The interest for these approaches was clearly manifested by the students deciding to follow these optional classes.

The making of this book owes much to the exchange with all, co-authors, reviewers or others, who were interested in fractal analysis and non-linear dynamics: Didier and Anne Sornette and Jean-Louis Cheminée for the applications of these methods to volcanic eruptions and the dynamics of eruptive systems; Claude Jaupart, Philippe Davy, Saint Clair-Dujonc, Jean-Louis Le Mouël, Adler Lomnitz, who discussed these works. The study of self-similarity in eruptive phenomena was conducted on the Anak-Krakatau complex with Christine Deplus and Michel Diament. Its demonstration on large sets of andesitic volcanoes results from a collaboration with François Chabaud and Eric Lewin. In the domain of rock fracturation and applications of the Cantor's dust technique, I owe much to the enthusiasm of Bruce Velde, as well as Béatrice Ledésert, G. Touchard and A. Badri. In geomagnetism, the data about field inversions, secular variations recorded in magnetic observatories or lake or seafloor cores, were the occasion for many exchanges and collaborations with Jean Besse, Jean-Pierre Valet, Vincent Courtillot, Yves Gallet, Gauthier Hulot, Jean-Louis Le Mouël. In seismology, I mentioned Lassaad Nouaili for his Ph.D., but seismic hazards were studied in collaboration between IPGP (Jean-Louis Le Mouël) and MIT-PAN in Moscow (Volodia Keilis-Borok), after formal agreements initiated by DRED (Claude Allègre, Vincent Courtillot). These agreements led to the exchange of scientists between the two institutes. Amongst those who played an important role in the introduction of fractal analysis and non-linear dynamics techniques, we must mention: Vladimir Pisarenko and his son Dimitri, wavelet transform expert, Andrei Gabrielov, Irina Rotwain, as well as Wladimir Kossobokov, Andrei Koklov, and Boris Bushkin, Peter Shebalin, Elena Blanter, Michael Schnirman, and Serguei Motlinski. These scientists did not limit themselves to applications in seismology, but also examined magnetic and telluric series and the theoretical modelling of the Earth's dynamo. In IPGP itself, collaborations were conducted with the members of the Geomagnetism Laboratory (Dominique Jault, Gauthier Hulot, Jean-Pierre Valet, Pierre Morat, Jean-Louis Le Mouël) and a few seismologists (Nicole Girardin, Pascal Bernard). To be thorough, I should mention the contribution of another group specialised in pattern recognition and Artificial Intelligence. Alexei Gvishiani and Mischa Zhizhin, theoreticians from the Institute of Earth Physics in Moscow, wrote the pattern recognition algorithm SPARS and brought us new techniques for computing the Lyapunov exponents. Finally, Alexei Panteleiev originated from the same institute, and as a post-doctoral researcher in our laboratory, increased the performance of our algorithm computing the correlation function (method of Grassberger and Procaccia) written by Claude Pambrun in 1988. He also looked at the fractal and multifractal analysis of the Earth's altimetric geoid.

In a domain further away from geophysics, I must mention the applications of the Cantor dust technique to palaeontology. Jean Chaline, Patrick Brunet-Lecomte and the scientists from the Laboratory of Vertebrate Paleontology in Dijon (France) provided data series which they interpreted in the context of evolution. From a methodology point of view, I owe a lot to a collaboration with Jacques Levy-Vehel, from INRIA, who started us in multifractal analysis, whose applications to long seismic, volcanic and geomagnetic series would reveal the interest.

The "club" founded by Daniel Shertzer in 1988 was a forum of exchange of ideas between different disciplines (first at the Météorologie Nationale, then at the Ecole Normale Supérieure, finally in Jussieu). Its regular meetings enabled the best following of the development of ideas.

I mention as well the lively discussions on the matter with my son Philippe, my daughter Judith, and Benoît Cibrario on epistemological grounds. Many readings and re-readings of this book allowed me to enhance the style and the content. First, I would like to thank Albert Tarantola, series editor at Masson, who spent a lot of time on this and suggested many revisions and ameliorations, and whose large editorial experience was very helpful. The group of students who followed the first lectures were asked to review the notes they had been given. Most did it diligently and thoroughly: first Lionel Hongre and Pascal Seilhac, who later worked in my laboratory, Philippe Thierry who realised the most minute analysis, Catherine Berge and Pascal Ultré-Guérard, as well as Raoul Beauduin. I would like to mention also Jacques Brassart, Rodolphe Cattin, Emmanuel Gauchez, Valérie Malavergne and Liêm Tran.

The practical realisation required many contributions. First, the shaping of the lecture notes which are the base of this book. It was accomplished by Catherine Netter, whom I would like to thank especially. She completely transcribed the first 200-page version, solving the many problems associated to formatting the document and the many equations. This document was rewritten in LaTeX. In this task, I was helped by Albert Tarantola, Geneviève Moguilny and Claude Mercier. The last two ones should be thanked for their remarkable patience.

I would like to associate them with Philippe Stoclet and Jeanine Mivielle. The former programmed and realised on the CM-2 the perspective views of the attractors for the secular variations of the geomagnetic field in sediments from Lake Bouchet, as well as the beautiful representation of quaternions. The latter achieved the prints and many of the figures before the final version of the manuscript. Their help was very important. I would like to mention that the figures were redrawn and translated into PostScript by Ion Cranana at IPG, and by John Butscher at ORSTOM-Nouméa. I am grateful to them for accomplishing this huge task. Jacques Deverchère gave me the beautiful recordings from the White Cordillera in Peru to illustrate the fractal analysis of seismicity, on which he worked for long hours during his Ph.D. in Orsay. During the final stage of this work, I benefited from the efficient help of Liêm Tran, of the Science Department from Masson. Liêm Tran is an ex-alumnus of the Geophysics M. Res., and followed the "Non-linear dynamics" option. This very favourable circumstance was extremely helpful, and I thank him vigorously.

Finally, it would be unforgivable to forget and not ask for their indulgence my close family, my wife Françoise Myriam and my children Philippe and Judith, who shared, maybe more than they would have wished for, the multiple worries associated with the writing of a book.

Homs, Aumessas, Paris
December 1991 - July 1993

List of Illustrations

Introduction
1	Fractionary Brownian model of topography	4
2	The length of Brittany's coastline	5
3	Burridge and Knopoff's model	6

Chapter 1
1.1	δ-covers of $\mathcal{F}$	12
1.2	Hausdorff dimension	14
1.3	Empirical estimate of the Hausdorff dimension	16
1.4	Hausdorff dimension - Box-counting	17
1.5	Triadic Cantor set and other Cantor sets	19
1.6	Random Cantor sets	22
1.7	Von Koch's curves	23
1.8	Classic fractal sets	24
1.9	Computation of the fractal dimension by box-counting	26
1.10	Projection of a fractal set	27
1.11	Cantor's Target	29
1.12	The harmonic oscillator	32
1.13	The simple pendulum	33
1.14	Magnetic pendulum with a central force field	34
1.15	Magnetic pendulum with a non-central force field	36
1.16	Area contraction for a pendulum kept oscillating	38
1.17	Sensitivity to Initial Conditions (SIC)	40
1.18	Phase trajectory in a three-dimensional space	41
1.19	The attractor's dimension	42
1.20	Relation of Atten and Caputo (1987)	43
1.21	Lorenz Attractor	45
1.22	Hénon's Attractor	46

List of Illustrations

1.23	Self-similarity of Hénon's attractor	47
1.24	Rössler's attractor	48
1.25	Linear and non-linear transforms.	49
1.26	Lyapunov exponents λ_1 and λ_2	51
1.27	Computation of the Lyapunov exponent	52
1.28	Method of the triangles' areas	52
1.29	Poincaré section	54
1.30	Poincaré section of the Lorenz Attractor	55
1.31	The quadratic map	57
1.32	A cascade of bifurcations	58
1.33	Notion of generalised dimension	61
1.34	Multifractal set built on a triadic Cantor set	64
1.35	Example of a singularity spectrum	65
1.36	Practical computation of a singularity spectrum	67
1.37	Wavelet analysis of Cantor sets with a Mexican Hat	69
1.38	Two wavelet transforms of a triadic Cantor set	71
1.39	Floquet matrix	73
1.40	Poincaré section of a phase trajectory	75
1.41	Poincaré section during a Hopf bifurcation	75
1.42	Theory of Ruelle and Takens	76
1.43	The Curry and Yorke model	78
1.44	Results of the map $\mathcal{F}$	79
1.45	The path to chaos	79

Chapter 2

2.1	Approximate coast lengths	82
2.2	Spectrum of a bathymetric profile south of the Azores	85
2.3	Spectral density spectrum	87
2.4	Cube fragmentation model	90
2.5	Fragmentation probability and dimension	91
2.6	Possible combinations during the fragmentation of a "sound pillar"	92
2.7	Model of the "sound pillar"	92
2.8	Graph (p_n, p_{n+1}) for the "sound pillar" model	93
2.9	Model of the "weakness plane"	93
2.10	Two-bump model	94
2.11	Self-similarity in fracturation	95
2.12	Fractal analysis of a fault field	96
2.13	Fractal analysis of a 2-D section from a fracture field	98
2.14	Cantor's dust method	98
2.15	Intersection of two sets made from fracture fields with perpendicular directions	100
2.16	Method of Aviles et al. (1987)	102
2.17	Method of Okubo and Aki (1987)	103
2.18	Percolation	105
2.19	Analogy with the percolation process	106
2.20	Examples of histograms (theoretical and for the eruptions of the Piton de la Fournaise between 1930 and 1994)	108
2.21	Cumulated distributions	110
2.22	Illustration of the Gutenberg-Richter law	112
2.23	Record of a seismic event in the White Cordillera, Peru	115

List of Illustrations   xxi

2.24	Fractal dimension of seismicity in Vanuatu	117
2.25	Fractal analysis using the Cantor dust technique	120
2.26	Application of the correlation function method	121
2.27	Example of a first-return map	122
2.28	First-return maps obtained with sub-series	123
2.29	First-return map for the eruptive series of Hawaii	124
2.30	Application of Fractal Singularities Analysis (FSA) to Hawaiian tremors	125
2.31	Self-organised criticality and the game of dominoes	127
2.32	Example of SOC on a sand heap	128
2.33	Results from the sand heap experiment	129
2.34	*A posteriori* results of the M8 algorithm in California	133
2.35	Construction of the CN algorithm	135
2.36	Results of the CN algorithm in California	136
2.37	Energy spectra for the periods of variation of the geomagnetic field	137
2.38	Model of the two-disk dynamo	138
2.39	Attractor of the dynamic system associated to Erschov's model	140
2.40	Inversions of the Earth's magnetic field	141
2.41	Application of the method of Grassberger and Procaccia (1983)	142
2.42	Variations of the Earth's magnetic field in the last 120,000 years	144
2.43	Dimensions of the attractors for the variations of the Earth's magnetic field	145
2.44	Analysis of the attractor for a core from Lake Bouchet	146
2.45	Perspective view of the attractor of the magnetic field variations	148

Chapter 3
| 3.1 | Error growth for local and polynomial predictors | 153 |
| 3.2 | Successive stages of analysis of a triadic Cantor set | 155 |

Chapter 4
4.1	Affine transform	161
4.2	Example of a series of affine transforms of an ivy leaf	162
4.3	Sierpinski's Triangle	163
4.4	Applications of the IFS code to the fern leaf	164
4.5	Fractal tree constructed with the IFS code	165
4.6	Example of Julia sets	166
4.7	Julia sets inside the Mandelbrot set	168
4.8	Sinaïs' billiards	169
4.9	The Baker's Transform	170
4.10	The "game of life"	171
4.11	Log-log graph of clusters and their duration in the "game of life"	172

Annex A
A.1	Example of the Venetian blind	180
A.2	Projection on a plane of geometric objects and fractal sets from $\mathcal{R}^3$	181
A.3	Poincaré section of Hénon's attractor	184
A.4	Poincaré section of a few trajectories	186

xxii List of Illustrations

A.5	Mastering chaos	187
A.6	Search for a stable orbit	187
A.7	Example of a cascade generating a binomial measure	190
A.8	Multifractal spectrum of the binomial measure from Figure A.7	191

Annex B

B.1	Geomorphology of a river network in the High Mbomou Basin	196
B.2	Fractal dimensions and hydrographic profiles	197
B.3	Application of the moving-window technique to a function defined on a spherical surface	200
B.4	Scale invariance of the altimetric geoid	202
B.5	Scaling law of the model of Allègre and Le Mouël (1994)	204
B.6	Breaking probabilities of a central square as a function of its environment	205
B.7	Fracturation in the self-healing and cooperating cases	208
B.8	Orientation of fractures at several levels	209
B.9	The SOFT model	210
B.10	Basic lattices and percolation	212
B.11	Degradation of a hexagonal lattice	213
B.12	Study of an infinite cluster at the percolation threshold	218
B.13	Shortest path between the extreme edges of a lattice	219
B.14	Space-time method of Shebalin and Rotwain	221
B.15	Time evolution of the function f for a seismic region of California	223
B.16	Energy spectrum of magnetic declination variations	224
B.17	Mutual information as a function of the time delay	226
B.18	Estimation of Lyapunov exponents for the Lorenz attractor	228
B.19	Mutual information and hourly magnetic declinations recorded at Chambon-la-Forêt	229
B.20	Graphical representation of an attractor with a high dimension	230
B.21	Estimation of Lyapunov exponents for the Chambon-la-Forêt data	232
B.22	Determinism in evolution	233

Colour plate section   (between pages 192 and 193)

1   Representation of a Mandelbrot set based on quaternions
2   Perspective view of a segment of the Mid-Atlantic Ridge near the Kane Fracture Zone
3   Analysis of the altimetric geoid
4   Representation in a pseudo-phase space of the secular variations of the Earth's magnetic field inclinations, recorded on cores from Lake Bouchet during the last 120,000 years

List of Tables

Chapter 2
2.1	Fractal dimensions observed when fragmenting various objects	89
2.2	Success rates of the M8 algorithm	132

Chapter 4
4.1	IFS code for Sierpinski's Triangle	163
4.2	IFS code for Sierpinski's Square	164
4.3	IFS code for the fern leaf	164
4.4	IFS code for the fractal tree	165

Annex B
B.1	Values of the percolation threshold for several lattice geometries	211
B.2	Values of the critical exponent as a function of the space dimension	216
B.3	Values of Lyapunov exponents for the time series of Chambon-la-Forêt	231

Introduction

> *Wild rabbits were wandering at all hours on the lawns, looking for earthworms which these animals are so fond of. Lagging behind them, their long tails were producing this characteristic screeching which explorers liked to recognise as perfectly innocuous.*
> Boris Vian, Vercoquin et le Plancton, 1966

1. NATURAL PHENOMENA AND NON-LINEAR DYNAMICS

The study of unordered environments has known a great success in physics for about twenty years, because they reflect particularly well most natural environments observed at a scale small enough. They can be studied either by trying to adapt the models used in ordered environments and supplementing them with new parameters, or by developing theoretical tools that account for this disorder. The techniques of fractal geometry, the study of non-linear dynamic systems, are using these new tools.

The pioneers in this recent evolution, Mandelbrot, Ruelle, Takens, Lorenz, Hénon, Thom, Kolmogorov, Prigogine, Keilis-Borok, Sinaï, have used the advances made by mathematicians at the end of the last century. First among them was Henri Poincaré whose ideas, according to Keilis-Borok, infiltrated themselves into all sciences one after the other, and always revolutionised them.

Benoît Mandelbrot's role has been primordial ever since he suggested in the 60s the study of the Earth, the sky, and the ocean with a large family of geometric objects judged until then esoteric and useless. The notion that leads him will be named with one of the two neologisms "fractal object" and "fractal", words that he coined out from the Latin *fractus* meaning irregular or broken. In 1967, following up the works of Richardson (1961), he acknowledges that the metrics used to measure coastal lengths makes it possible to introduce the definition of a dimension. This definition was already established in pure mathematics but no one had thought it could also be applied to practical matters.

This need to explore unordered environments expressed itself in many domains. About the processing of numerical series, René Thom wrote: "*It is clear that the theory of Fourier series really nspired itself from Physics, more specifically from the study of vibrating strings or from heat theory; this is the reason why I came to slightly abhor this theory; it is one of these algorithms that most mathematicians and scientists in general adore; but for me, I always found it irritatingly linear. Too linear to be serious !*"

This infiltration in all sciences was underlined by David Ruelle and Ilya Prigogine. The former remarked that the similarity of complex behaviours in hydrodynamics, chemical kinetics, mechanics or electronics, concerns experimental details made accessible only recently. This similarity results from a modern theory of non-linear systems, or, more precisely, from the qualitative theory of differentiable dynamic systems. Prigogine et al. (1991) resumed the study of LPS (Large Poincaré Systems), which form a large class of non-integrable systems classed by Poincaré (1892). In such a system, the study of time variations of an observable generally shows a continuous spectrum, which is one of the characteristics of chaos. These LPS are characterised by a continuous spectrum and the presence of continuous resonance sets. And it is well known that non-integrability leads to the apparition of random trajectories and chaos. This non-integrability is due to the resonances responsible for divergences.

The mathematician Paul Levy (in Mandelbrot, 1975) was acknowledging, on the subject of the famous von Koch curve, that it was in some way looking like a rocky coast: "*Without doubt, our intuition was forecasting that the absence of tangent and the infinite coast length are linked to details so small they cannot be tracked down. I insist on this role of intuition as I was always surprised to hear people say that geometric intuition fatally led to think that every continuous function was derivable. Ever since my first encounter with the notion of derivative, my personal experience had taught me the opposite.*"

Among the principal concepts called upon in this work, those of measure and dimension play a very important role in non-linear dynamics, and a chapter inspired from the work of Falconer (1990) is devoted to them. They shall be found again in the next chapter treating of multifractal sets, which will lead to precise and generalise these notions: Lebesgue metrics, Hausdorff metrics, Hausdorff dimension, Hausdorff-Besicovitch dimension, similarity dimension or Mandelbrot's fractal dimension, information dimension, correlation dimensions of order 2, 3, 4 ..., Renyi dimension or generalised dimension.

The definitions and properties of these concepts result from works by many mathematicians, among them Minkowski (1901), Lebesgue (1918), Hausdorff (1919), Bouligand (1928), Pontrjagin and Schnirelman (1932), Besicovitch (1935, 1939), Kahane (1971, 1976). The geometrical aspect of objects or "*pathologic*" curves evokes the names of mathematicians such as Cantor, Peano, von Koch, Sierpinski and Poincaré.

If, very roughly, dimension describes part of the space occupied by a set investigated at a very small scale (it brings information about its geometrical properties), the measure is a more complicated concept illustrated by mass or probability distributions; two sets may well be identical (through translation) although having different masses. Therefore the transition from fractal sets to multifractal measurements implies the need to specify the metrics distribution function; whereas a fractal set needed only one fractal dimension, this number is replaced by a function (Frisch and Parisi, 1985).

In this quest for a more and more precise description of fractal sets, it is easy to see that progress still needs to be made. Along the last ten years, a great effort was dedicated to the characterisation of measures supported by fractal sets. This led to multifractals, but generalised fractal dimensions and the singularities spectrum only provide statistical information about the respective contributions of each singularity. To learn more and get the information about the spatial location of these singularities, fractal functions were analysed with wavelet techniques. These wavelet transformations form a tool close to a "mathematical microscope" which makes it possible to describe the hierarchy underlying the structural complexity of multifractal objects.

Before giving the plan of this work, let us remark that the study of fractal objects very early showed the aesthetic aspect of figures generated with more and more powerful computers (plate 1). It is possible to epilogue about this incursion into art; this could be considered as the simplified and purified translation of the "harmonies of nature" amidst which we live and where fractal laws are dominant (topography, flows, eddies, diffusion, fragmentations, crystallisations, etc.). These drifts probably were prejudicial to the development of the new ideas at the basis of fractal research. The first approaches were and still are resisted by the traditional resistance of the scientific community (which thinks of them as a passing fashion or a "gadget" !). But this kind of reactions was studied enough epistemologically, for example in the excellent book from Thomas Kuhn (1970), and there is no need to talk further about it.

The field of applications is very large, because many natural phenomena exhibit a chaotic dynamics. This idea, wrote Keilis-Borok (1990), was inaugurated by Henri Poincaré and penetrated all areas of science, one after the other, and revolutionised them.

2. OUTLINE OF THIS WORK

In this study, the generalities and theoretical notions will introduce the notions of measure and dimension before reaching the very large domain of fractal sets, which will be limited to the essential.

Dynamic systems coming from non-integrable differential systems appeared to us as very rich for their applications to the non-linear dynamics of geophysical processes. The definitions and main properties of phase spaces, phases trajectories, strange (or chaotic) attractors, and Lyapunov exponents are based on the work of Bergé et al. (1984) and Falconer (1990), and on articles from a CNRS colloquium about the numerical processing of strange attractors (Cosnard, 1987). The applications of the first return, the processing of numerical series, multifractals, wavelet transforms, stem from compilations of articles published in specialist journals about Physics: *Physica D, Physical Review Letters, Physics Letters, Physical Review* etc., and stem as well from

4   Introduction

personal research in the fields of volcanology, seismology, geomorphology and geomagnetism.

Self-Organised Criticality (SOC) encounters a large success in many domains and will not be detailed in this general section but in the following one, because of its important applications in seismology. SOC considers dynamic systems in their critical state, permanently self-organising; a continuous input, of sand grains in the classical model of the sand heap, of continuous stress and deformation in the case of a plate boundary, leads in a chaotic and deterministic fashion to catastrophic events of discharge or stress release whose amplitude and timing obey power laws and are self-similar.

The objective of this work being non-linear dynamics in geophysics, an important part of this book will deal with the many applications that have been developed in this field during the last 10 years.

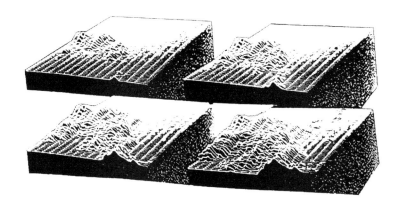

Figure 1. Fractionary Brownian model of topography. The parameters were chosen so that the dimensions of the coasts and vertical profiles are equal to 1.3. The dimension of the surface is therefore 2.3. From Mandelbrot (1975).

Historically, the study of nature's shapes has been the aim of the first works from several pioneers (Mandelbrot, Thom). The similarity with natural reliefs on the Earth, at the bottom of the oceans, on planets, were very soon noticed with Mandelbrot's fractionary Brownian model. An example is given in Figure 1. A first chapter will be dedicated to geomorphology where power laws work well. Let us remember that the concept's introduction stems from the study of the length of British coasts by Richardson (1961) (Figure 2). The spectral methods previously used are easily comparable with fractal analysis. The richness of fractal analyses of the seafloor will be insisted upon, going back to the mechanism generating topography.

The study of rock fragmentation and fracturation grew rapidly after the first study by Allègre et al. (1982). A pitfall related to the use of renormalisation groups slowed the modelling, but recent works linked with seismology broadened the field of applications.

The description of faults and fracture fields has always been a major preoccupation of tectonicians and structuralists. The fractal tool found there a privileged niche and it probably is in Earth sciences the domain where "fractal" scientific production has been predominant. From the study of downscaled analogue models through experiments on rocks and up to field data, the activity is important, and past and current approaches will be extensively reviewed. Self-similarity of the fractal process was demonstrated on a granite dome to span five orders of magnitude (Velde et al., 1990). The chapter ends by examining the percolation process, privileged domain of solid physics and physics of condensed matter, for which the non-linear approach dates back to the 70s (de Gennes, 1976). Applications to petroleum geology, geothermics, hydrology, fractured environments, and their civil engineering applications are vast and the main ones will be explained.

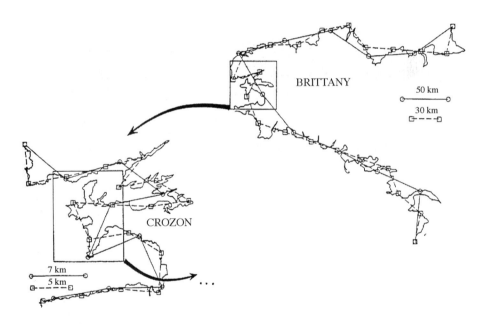

Figure 2. Brittany's coast length. Example of measuring a coastline cut down with Richardson's method; Brittany's coast is surveyed with steps of decreasing lengths.

Seismology played an important role in the development of fractal ideas since the analogy was recognised between the famous law of Gutenberg and Richter (1949) and power laws from fractal geometry (Aki, 1980). The already old studies about the parameter b (Log.N= $a + b$ M), the introduction of fractal models and chaotic dynamic systems extended the description of focal mechanisms. The non-linear character was also recognised for some models like Burridge and Knopoff's (1967, Figure 3), which belongs perfectly to the SOC framework proposed 20 years later by Bak et al. (1987). A chapter is devoted to the law of Gutenberg and Richter and its fractal implications, and to the link with renormalisation groups and fragmentation as it was established by Smalley et al. (1987).

Another aspect of the fractal approach is its probabilistic implications. Mandelbrot (1975) established the relation between the probability of occurrence of intermissions (time lapse between two successive events) when they are distributed along a power law $Pr(U > u) = u^{-D}$, which reads as : the probability that, immediately after an event (earthquake, eruption, etc.), there is a time lapse greater than or equal to u is u^{-D}, where D is the dimension of the set of time series of past intermissions and events. This probabilistic approach was applied by Smalley et al. (1987). The extension to a multidimensional space was developed in parallel by the Russian school which pairs it with pattern recognition techniques (Keilis-Borok and Kossobokov,1986).

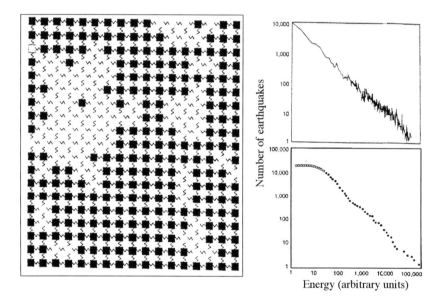

Figure 3. Burridge and Knopoff's model (Burridge and Knopoff, 1967). Occurrences of earthquakes from all sizes can be simulated. Re-examined 20 years later by Bak and Chen in an SOC optics, this model gives interesting descriptions of earthquakes clustering in time and space. From Bak and Chen, 1991.

Two chapters deal with this approach and regroup earthquakes and volcanic eruptions. The tests used in the processing of time series call upon three different types of approaches explained in the theoretical section: the analysis of fractal sets by Cantor's dust method; the study of the attractor for the dynamic system generating the series by Grassberger and Procaccia's method (1983); the first-return map and the properties of 2-D and 3-D recurrences.

For the series of earthquakes and volcanic eruptions, it will be demonstrated that Poissonian models generally used are faring less well than the power law distributions. Thus we will bring into evidence the existence of deterministic systems with few degrees of freedom. These first results show that the techniques used are particularly well adapted to this type of natural processes. A more thorough modellisation with SOC gives glimpses of new possibilities, even if, in the current state of the theory, the "seismic flow" or the "eruptive flow" cannot be clearly and entirely formulated. A chapter is devoted to this field, where we develop the prediction methods of the Russian school by detailing the algorithms M8 (prevision of earthquakes with magnitudes greater than 8) and CN (previsions in the California-Nevada region).

Among the privileged domains to which non-linear dynamics seems particularly suited, one must include the variations of the Earth's magnetic field. For inversion phenomena in apparently chaotic series (McFadden and Merrill, 1984), or for secular variations, the similarities between records and intermittence processes is striking. The studies undertaken in this domain were realised along two ways. The first made use of Rikitake's famous model of self-maintained dynamo, for which Ershov et al. (1989) show an attractor with a dimension varying between 2.09 and 3.06, according to the ohmic dissipation coefficient chosen. This chaotic deterministic approach is as visible with the Earth's magnetic field: time inversion series (Dubois and Pambrun, 1990) and time series of the field vector as recorded, for example, in marine and lacustrine sediments. Two examples of the study of time series will be presented and the deterministic character of the geomagnetic field's variations will be demonstrated.

It is altogether clear that the self-imposed limits of the present work are arbitrary. There are many other domains where the analysis techniques presented here could also be applied: gravimetry, electromagnetism, propagation of seismic waves in heterogeneous environments, focal mechanisms, mantle convection, or, in traditional geology, sedimentology, palaeontology and evolution, hydrogeology, clinology, etc. These perspectives will be touched upon in a chapter which underlines how many natural processes can be analysed this way.

Finally, the "fun" aspect of fractals, strange attractors and dynamic systems in a critical state should not be ignored because it has played an important part in the propagation of these ideas. It is therefore dealt with in an additional chapter.

The Annexes hold everything that might be more interesting to a reader more keen about a specific subject. Its technical aspect did not seem fit for the main body of a book of introduction to non-linear techniques aimed at post-graduate and doctoral students.

During the time of writing and publishing, many articles have been added to those used in this book, and it is now impossible to reference them all as it was possible in 1991. This demonstrates that, little by little, these approaches are getting recognised by the geophysics community.

1

Fractal and Multifractal Theory

> *Things that coincide are equal*
> Euclid, Common Notions, Book 1

1.1 NOTION OF MEASURE

By definition, measuring simply consists in giving a numerical value (or size) to a set, so that, if this set is decomposed into a finite and countable number of subsets, the value of the whole is the sum of the subsets' values. The next definitions can be supplemented, more rigorously, by the reading of Kahane (1976) and Falconer (1990), as well as that of Annex A.1. Readers unfamiliar with the subject may go directly to page 15.

μ is a measure on $\mathcal{R}^n$ if each subset of ^{n} can be attributed a positive (and possibly infinite) number so that:

- if $\emptyset$ is the null set, $\quad \mu(\emptyset) = 0$ \hfill (1.1)

- if $\mathcal{A} \subseteq \mathcal{B}$, $\quad \mu(\mathcal{A}) \leq \mu(\mathcal{B})$ \hfill (1.2)

- if $\mathcal{A}_1, \mathcal{A}_2, \ldots$ is a countable series of sets,

$$\mu\left(\bigcup_{i=1}^{\infty} \mathcal{A}_i\right) \leq \sum_{i=1}^{\infty} \mu(\mathcal{A}_i) \qquad (1.3)$$

or, if the A_i are disjoint Borelian sets[1],

$$\mu\left(\bigcup_{i=1}^{\infty} A_i\right) = \sum_{i=1}^{\infty} \mu(A_i) \qquad (1.4)$$

$\mu(A)$ is called the measure of the set A and $\mu(A)$ is considered as the value (or size) of A measured in a certain way.

Equations (1.1) to (1.4) enable one to define a measure over $\mathcal{R}^n$ at three conditions:
1. The measure of a null set is zero (equation 1.1)
2. The larger a set, the larger its measure (relation 2.2)
3. If a set is the union of a countable number of parts (possibly overlapping), the sum of the measures of the parts is at least equal to the measure of the whole (relation 1.3). If the parts do not overlap (e.g. Borelian sets), then the sum of the measures of the parts equals the measure of the whole (equation 1.4).

This leads us to a few examples of common measures.

1.1.1 Measuring by counting

For each set A of $\mathcal{R}^n$, let $\mu(A)$ be the number of points in A. If A is finite, then $\mu(A)$ is a measure of A on $\mathcal{R}^n$.

1.1.2 Mass of a point

Let a be a point of $\mathcal{R}^n$, and by definition let $\mu(A)$ equal 1 if a is in A, and 0 if not. Then μ is a mass distribution, taken as the mass of a point centred in a.

1.1.3 Lebesgue measure in $\mathcal{R}^1$ and $\mathcal{R}^n$

The Lebesgue measure $\mathcal{L}^1$ extends the notion of length to a large collection of subsets of $\mathcal{R}$, including the Borelian sets. One defines :

---

[1] The class of Borelian sets is the smallest collection of subsets of $\mathcal{R}^n$ which have the following properties :
- every open set and every closed set is a Borelian set
- the union of any finite and denumerable collection of Borelian sets is itself a Borelian set, and the intersection of any finite and denumerable collection of Borelian sets is a Borelian set.

$$\mathcal{L}^1(a,b) = \mathcal{L}^1[a,b] = b - a \tag{1.5}$$

If $\mathcal{A} = \bigcup_i [a_i, b_i]$ is the finite and countable union of the disjoint intervals (indices i), $\mathcal{L}^1(\mathcal{A}) = \sum_i (b_i - a_i)$ is called the length of $\mathcal{A}$, i.e. the sum of the interval's lengths. This brings in the definition of the Lebesgue measure $\mathcal{L}^1(\mathcal{A})$ of a set $\mathcal{A}$:

$$\mathcal{L}^1(\mathcal{A}) = \inf\left\{ \sum_{i=1}^{\infty} (b_i - a_i) \,;\, \mathcal{A} \subset \bigcup_{i=1}^{\infty} [a_i ; b_i] \right\} \tag{1.6}$$

This means that we have to look for all elements covering $\mathcal{A}$ by a countable series of intervals, and that we take the smallest total of interval lengths. The generalisation to $\mathcal{R}^n$ gives the measure of Lebesgue $\mathcal{L}^n(\mathcal{A})$ on $\mathcal{R}^n$.

If $\mathcal{A} = \{(x_1, \ldots, x_n) \in \mathcal{R}^n;\, a_i \le x_i \le b_i\}$ is a parallepiped in $\mathcal{R}^n$, the n-dimensional volume of $\mathcal{A}$ is : $\mathrm{vol}^n(\mathcal{A}) = (b_1 - a_1)(b_2 - a_2) \ldots (b_n - a_n)$.

The n-dimensional Lebesgue measure can be considered as the extension of the n-dimensional value to a series of sets. By analogy to $\mathcal{L}^1(\mathcal{A})$, $\mathcal{L}^n(\mathcal{A})$ will be defined as:

$$\mathcal{L}^n(\mathcal{A}) = \inf\left\{ \sum_{i=1}^{\infty} \mathrm{vol}^n(\mathcal{A}_i) \,;\, \mathcal{A} \subset \bigcup_{i=1}^{\infty} \mathcal{A}_i \right\} \tag{1.7}$$

in which the infimum will be taken among all elements covering $\mathcal{A}$ with parallepipeds $\mathcal{A}_i$ (instead of intervals). It is clear that $\mathcal{L}^n(\mathcal{A}) = \mathrm{vol}^n(\mathcal{A})$ if $\mathcal{A}$ is a parallelepiped or any set for which the volume can be determined by the usual measuring rules.

1.1.4 Hausdorff Measure

Let $\mathcal{U}$ be any non-empty set of the n-dimensional Euclidean space $\mathcal{R}^n$. The diameter of $\mathcal{U}$ is :

$$|\mathcal{U}| = \sup\left\{ |x - y|;\, x \in \mathcal{U};\, y \in \mathcal{U} \right\} \tag{1.8}$$

i.e. the largest distance between any pair of points in $\mathcal{U}$. If $\{\mathcal{U}_i\}$ is a series of sets covering $\mathcal{F}$ with diameters of δ at most,

$$\mathcal{F} \subset \bigcup_{i=1}^{\infty} \mathcal{U}_i, \quad 0 < |\mathcal{U}_i| \leq \delta \text{ (for any } i\text{)} \tag{1.9}$$

$\{\mathcal{U}_i\}$ is called a δ-cover of $\mathcal{F}$. If $\mathcal{F}$ is a subset of $\mathcal{R}^n$ and s is a positive number, the Hausdorff measure is defined for each $\delta > 0$ as :

$$\mathcal{H}_\delta^s(\mathcal{F}) = \inf.\left\{ \sum_{i=1}^{\infty} |\mathcal{U}_i|^s \, ; \, \mathcal{U}_i \text{ being a } \delta\text{-cover of } \mathcal{F} \right\} \tag{1.10}$$

This is the minimum of the sum of the diameters of the δ-cover of $\mathcal{F}$ to the power s (Figure 1.1).

Figure 1.1. A set $\mathcal{F}$ and two possible covers by elements δ. The minimum of $\sum_i |u_i|^s$ on all δ-covers of the $\{u_i\}$ gives $\mathcal{H}_\delta^s$. From Falconer (1990).

When δ decreases, the number of possible covers of $\mathcal{F}$ diminishes. Therefore, the value of $\mathcal{H}_\delta^s(\mathcal{F})$ increases and tends toward a limit when δ is closer to zero:

$$\mathcal{H}_\delta^s(\mathcal{F}) = \lim_{\delta \to 0} \mathcal{H}_\delta^s(\mathcal{F}) \tag{1.11}$$

This limit exists for any set $\mathcal{F}$ of $\mathcal{R}^n$, and $\mathcal{H}^s(\mathcal{F})$ is called the Hausdorff measure of $\mathcal{F}$ with the dimension s. It is possible to demonstrate that $\mathcal{H}^s(\mathcal{F})$ is indeed a measure, as in the beginning of this chapter. It is also possible to demonstrate that, for subsets of $\mathcal{R}^n$, the Hausdorff measure with n dimensions is the Lebesgue measure with n dimensions, at a constant C_n apart.

More precisely, if $\mathcal{F}$ is a Borelian subset of $\mathcal{R}^n$, then : $\mathcal{H}^s(\mathcal{F}) = C_n \, vol^n(\mathcal{F})$

If the dimension n is even (n = 2m, m ∈ $\mathcal{N}$, $C_n = C_{2m} = \dfrac{\pi^m}{m!}$

If the dimension n is odd (n = 2m+1, m ∈ $\mathcal{N}$, $C_n = C_{2m+1} = 2^{m+1} \dfrac{\pi^m}{(2m+1)!!}$

This constant C_n is the volume of a sphere of diametre 1.

For common subsets of $\mathcal{R}^n$, $\mathcal{H}^0(\mathcal{F})$ is the number of points in $\mathcal{R}^n$; $\mathcal{H}^1(\mathcal{F})$ is the length of a continuous curve of $\mathcal{F}$; $\mathcal{H}^2(\mathcal{F}) = \pi/4 \times$ area($\mathcal{F}$) if $\mathcal{F}$ is continuous surface; $\mathcal{H}^3(\mathcal{F}) = 4\pi/3 \times$ vol.($\mathcal{F}$) and $\mathcal{H}^n(\mathcal{F}) = C_n \, vol^n(\mathcal{F})$, if $\mathcal{F}$ is a volumic subset with n dimensions.

Let us note that the appearance of π in the relation between the Lebesgue and Hausdorff measures is due to the differences in their modes of computation: covering by n-dimensional parallelepipeds for the Lebesgue measure, and covering by spheres of diameter δ for the Hausdorff measure.

1.2 NOTION OF DIMENSION

The term of dimension, which we introduced above, was characterising Euclidean spaces in which we had established the notion of measure. In this section, we will extend this definition to non-integer values, which will allow the introduction of new objects, the *fractal sets*.

The difference between the Lebesgue dimension and the Hausdorff dimension comes from the fact that the Hausdorff measure with s dimensions is a generalisation of the Lebesgue measure to a dimension which does not need to be an integer. In equation (1.7), n is an integer, whereas in equation (1.10), s need not be.

1.2.1 Hausdorff Dimension

Let us come back to the definition of the Hausdorff measure:

$$\mathcal{H}^s_\delta(\mathcal{F}) = \inf. \left\{ \sum_{i=1}^{\infty} |\mathcal{U}_i|^s \, ; \, \mathcal{U}_i \text{ being a } \delta\text{-cover of } \mathcal{F} \right\} \quad (1.12)$$

It appears clearly that for a given set $\mathcal{F}$ and $\delta < 1$, $\mathcal{H}^s_\delta(\mathcal{F})$ does not increase with s. Thus, according to the relation $\mathcal{H}^s(\mathcal{F}) = \lim\limits_{\delta \to 0} \mathcal{H}^s_\delta(\mathcal{F})$, $\mathcal{H}^s(\mathcal{F})$ is not an increasing function. If $t > s$ and $\mathcal{U}_i$ is a δ-cover of $\mathcal{F}$,

$$\sum_{i=1}^{\infty} |U_i|^t \leq \delta^{t-s} \sum_{i=1}^{\infty} |U_i|^s \qquad (1.13)$$

And by considering the infimum of each term :

$$\mathcal{H}_\delta^t(\mathcal{F}) \leq \delta^{t-s} \mathcal{H}_\delta^s(\mathcal{F}) \qquad (1.14)$$

If $\delta \to 0$ and $\mathcal{H}^s(\mathcal{F}) < \infty$, then $\mathcal{H}^t(\mathcal{F}) = 0$ for $t > s$.

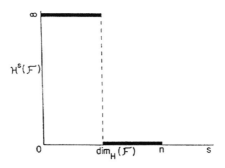

Figure 1.2. Hausdorff dimension. Graph of $\mathcal{H}^s(\mathcal{F})$ as a function of s, for a given set $\mathcal{F}$. The Hausdorff dimension is the value of s where the jump from ∞ to 0 occurs. From Falconer, 1990.

The graph of $\mathcal{H}^s(\mathcal{F})$ as a function of s therefore shows a critical value for which $\mathcal{H}^s(\mathcal{F})$ switches from ∞ to 0. This critical value is called the Hausdorff dimension of $\mathcal{F}$, and is noted $\dim_H(\mathcal{F})$. Sometimes, it is also called the Hausdorff-Besicovitch dimension.
Thus:

$$\dim_H(\mathcal{F}) = \inf\left\{s;\ \mathcal{H}^s(\mathcal{F}) = 0\right\} = \sup\left\{s;\ \mathcal{H}^s(\mathcal{F}) = \infty\right\} \qquad (1.15)$$

so that (cf. Figure 1.2):

$$\mathcal{H}^s(\mathcal{F}) = \begin{cases} \infty & \text{if } s < \dim_H(\mathcal{F}) \\ 0 & \text{if } s > \dim_H(\mathcal{F}) \end{cases} \qquad (1.16)$$

Notion of Dimension

If $s = \dim_H(\mathcal{F})$, $\mathcal{H}^s(\mathcal{F})$ may be zero or infinite. It can satisfy: $0 < \mathcal{H}^s(\mathcal{F}) < \infty$.

If we look for example at the Hausdorff dimension of a flat disc in $\mathcal{R}^3$, according to the properties of length, surface and volume, we have:

$$\begin{cases} \mathcal{H}^s(\mathcal{F}) = \text{length}(\mathcal{F}) = \infty \\ 0 < \mathcal{H}^2(\mathcal{F}) = \pi/4 \times \text{area}(\mathcal{F}) < \infty \\ \mathcal{H}^3(\mathcal{F}) = 4\pi/3 \times \text{vol.}(\mathcal{F}) = 0 \end{cases} \quad (1.17)$$

because the disc is infinitely flat.

Thus $\dim_H(\mathcal{F}) = 2$, with: $\mathcal{H}^s(\mathcal{F}) = \infty$ if $s < 2$ and $\mathcal{H}^s(\mathcal{F}) = 0$ if $s > 2$.

The Hausdorff dimension of a disc is an integer, and equals its Euclidean dimension. But it should remain clear that the Hausdorff dimension of some sets may not be an integer. For example, the triadic Cantor set has a dimension comprised between 0 (Euclidean dimension of a point) and 1 (Euclidean dimension of a filled segment).

1.2.2 Another Definition of the Hausdorff Dimension

A fundamental idea, which recurs in most definitions of dimension, is linked to the notion of measure with a step δ.

For each δ, the set is measured ignoring the irregularities smaller than δ. Let us examine how these measures are affected when $\delta \to 0$.

For example, if $\mathcal{F}$ is a plane curve, the measure $M_\delta(\mathcal{F})$ may be the number of steps necessary for a pair of points spaced apart from δ to cover $\mathcal{F}$. A well-known example is given in Richardson (1939) and in Mandelbrot (1967), measuring the length of Brittany's coast with a compass whose tips are spread from δ.

Let us suppose that the measure $M_\delta(\mathcal{F})$ follows a power law in δ when δ tends towards zero, i.e. $M_\delta(\mathcal{F}) \approx c\delta^{-s}$, c and s being constant. According to definition (1.10), it is possible to say that $\mathcal{F}$ has the dimension s and that c is the "length of dimension s" of $\mathcal{F}$.

Using logarithms:

$$\log M_\delta(\mathcal{F}) = \log c - s \log \delta \quad (1.18)$$

The log-log graph in Figure 1.3 shows successive measurements of $\mathcal{F}$ (for example a coast length measured with a step δ) as a function of δ (δ being small and tending toward zero). If these points are aligned, it means that equation (1.18) is verified, and $M_\delta(\mathcal{F}) \sim c\delta^{-1}$. According to definitions, the dimension of $\mathcal{F}$ is s and its measure is c.

$$s = \lim_{\delta \to 0} \frac{\log M_\delta(\mathcal{F})}{-\log \delta} \quad (1.19)$$

This formula can be applied to experimental data, and s can be estimated as the gradient of the log-log graph plotted for a reasonable number of values of δ (Figure 1.3).

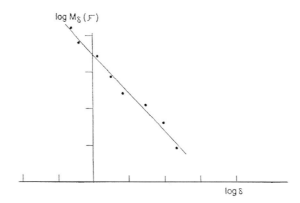

Figure 1.3. Hausdorff Dimension. Empirical estimate of the dimension of the set $\mathcal{F}$ on the log-log graph of $M_\delta(\mathcal{F}) \sim c\delta^{-s}$, where $M_\delta(\mathcal{F})$ is the number of steps δ necessary to cover the whole of $\mathcal{F}$.

This experimental definition comes back to the simplified definition of the Hausdorff dimension given in most works (i.e. Bergé et al., 1984) (Figure 1.4) :

$$D = \lim_{\varepsilon \to 0} \frac{\log. N(\varepsilon)}{\log.(1/\varepsilon)} \qquad (1.20)$$

where $N(\varepsilon)$ is the minimum number of cubes of side ε necessary to cover the whole of $\mathcal{F}$. If, for a small ε, this number $N(\varepsilon)$ varies as ε^{-D}, then D is the dimension of $\mathcal{F}$.

In the case of a single point, $N(\varepsilon) = \text{cst} = 1$, for any value of ε, and therefore $D = 0$. The Hausdorff-Besicovitch dimension of a point is 0, like its Euclidean dimension.

If $\mathcal{F}$ regroups all points belonging to a segment of length L,

$$N(\varepsilon) = \frac{L}{\varepsilon} = L\,\varepsilon^{-1} \qquad (1.21)$$

The dimension will therefore be $D = 1$, identical to the Euclidean dimension.

The Cantor dust is the perfect triadic set (see Section 1.3.3), and its Hausdorff dimension verifies: $0 < D < 1$ (Figure 1.5). Indeed, for $\varepsilon = 1/3$, the number of segments needed to cover the whole object is $N(1/3) = 2$. At the next iteration, $\varepsilon = 1/9$ and $N(1/9) = 4$, etc. At the m-th iteration, $\varepsilon = (1/3)^m$ and $N(\varepsilon) = N((1/3)^m) = 2^m$.

Therefore, according to equation (1.20):

$$D = \lim_{m \to 0} \frac{\log(2^m)}{\log(3^m)} = \frac{\log 2}{\log 3} = 0.6309 \qquad (1.22)$$

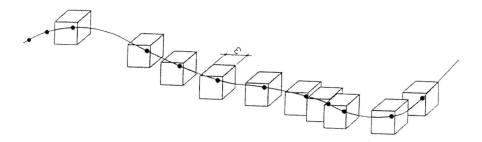

Figure 1.4. Hausdorff Dimension. Example of a way to cover a set of points with cubes of length ε. From Bergé et al. (1984).

1.3 FRACTAL SETS

> *But the mathematician does not discover, he invents*
> L. Wittgenstein, Remarks about the bases of mathematics, 1937

1.3.1 Definitions

The notion of fractal set was introduced early on during the definitions of measures and dimensions. The term of "fractal" was first employed by Mandelbrot (1967), who defines it as an object with a Hausdorff dimension strictly greater than its topological dimension. Before Mandelbrot's studies, the sets or functions that were not continuous or regular enough were considered as "pathological" and worth studying. They were regarded as oddities and rarely as a class to which a mathematical process could be applied (Cantor, Peano, Levy, Poincaré, etc.). This behaviour changed in the last years. It was realised that many properties could be brought into evidence in the mathematical domain of irregular (non-continuous) sets. Moreover, irregular sets allow to represent and describe natural phenomena much better than is possible with the objects of classical geometry. Fractal geometry provides a general framework to the study of these irregular sets because, although some cases are very regular or supposed to be, Euclidean geometry is unable to account at a macroscopic scale for the disordered structure of such natural objects as clouds, mountain ranges, or sea shores (Grunberger, 1991).

Falconer (1990) gives an interesting analysis of this concept: "*My personal feeling is that the definition of the fractal concept could be envisioned in the same way than a biologist thinks of the definition of the concept of 'life'. There is no robust and simple definition, but rather a list of properties characteristic of a living being, such as the ability to reproduce, or to move, or to exist. Most living beings possess most of the list's characteristics, despite some exceptions. In the same way, it seems more correct to consider a fractal as a set with the properties defined below, rather than limit ourselves to a precise definition which would certainly exclude some interesting cases*".

When referring to a fractal set $\mathcal{F}$, the following properties must be kept in mind:
1. The set $\mathcal{F}$ has a fine structure, such as, for example, details at arbitrarily small scales.
2. $\mathcal{F}$ is too irregular to be described with traditional words, locally as well as globally.
3. $\mathcal{F}$ often presents some kind of internal homothety.
4. Defined in one way or another, the *fractal dimension* of $\mathcal{F}$ is greater than its topological dimension.
5. In most cases, $\mathcal{F}$ is defined in a simple, often recursive, way.

Practically speaking, a fractal set will have the following property (Mandelbrot, 1975; Huang and Turcotte, 1989):

$$N_i = C / r_i^D \qquad (1.23)$$

where N_i is the number of objects of linear dimension r_i making up the whole set; D is the fractal dimension, and C is a proportionality constant.

For a continuous distribution, the generalisation of this equation is:

$$N = C / r^D \qquad (1.24)$$

where N is the number of objects making up the whole set, whose linear dimension is greater than r.

Many physical processes can be found to satisfy this relation. Examples in Earth sciences include the Korcak relation for the number of islands with areas greater than a specified value (Mandelbrot, 1975); the Rosin law for the number-size distribution of rock fragments (Turcotte, 1986), or the Gutenberg-Richter relation between frequency and magnitude of earthquakes (Aki, 1981). These applications to Earth sciences and to geophysics will be more detailed in Sections 2.1, 2.2 and 2.3.

1.3.2 Triadic Cantor Set

The starting point is the full segment ε_0, corresponding to the interval [0 ; 1] (Figure 1.5). ε_1 is the set obtained by removing the median third, so that ε_1 is made of the two intervals [0 ; 1/3] and [2/3 ; 1]. ε_2 is obtained by removing the median third of each non-empty interval of ε_1. ε_2 will therefore be made of the 4 intervals [0 ; 1/9], [2/9 ; 3/9], [6/9 ; 7/9]; [8/9 ; 1]. Proceeding further and further, ε_k is obtained by removing the median third of each interval of ε_{k-1}. ε_k will therefore be made of 2^k intervals, each of length 3^{-k}.

Mathematically, the Cantor set is the intersection $\bigcap_{k=0}^{\infty} \varepsilon_k$.

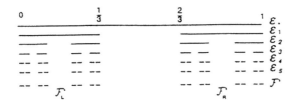

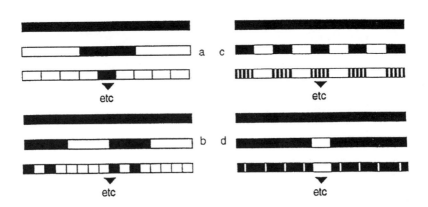

Figure 1.5. (top) Triadic Cantor set. $\mathcal{F}$ is obtained by removing the median third of intervals. (bottom) Cantor sets generated differently. The values of D are : (a) D = 0 (the set is reduced to one point); (b) D = log.2 / log.4 = 0.5; (c) D = log.5 / log.9 = 0.732; (d) D = log.8 / log.9 = 0.946. From Dubois and Cheminée, 1988.

It can be considered as the limit of the series of ε_k when k tends toward ∞ (Figure 1.5). The "dust" made from segments shortening with each iteration will be finer and finer. Very quickly and at a relatively small value of k, the "materiality" of the set inside the interval [0 ; 1] becomes quite tenuous, close to nothing. Hence the name of "Cantor Dust" often given to this set.

The mode of partition of the interval [0 ; 1] can of course be changed (cf. Figure 1.5). The resulting sets can then be described, quantified, by the attribution of one value, their dimension.

1.3.3 Dimension of the Triadic Cantor Set

Keeping the previous section in mind, let us compute the Hausdorff dimension. The triadic Cantor set $\mathcal{F}$ is divided into a left part: $\mathcal{F}_L = \mathcal{F} \cap [0; 1/3]$ and into a right part: $\mathcal{F}_R = \mathcal{F} \cap [2/3 ; 1]$. These two parts are of course geometrically similar to $\mathcal{F}$, but with a 1/3 ratio, and $\mathcal{F} = \mathcal{F}_L \cup \mathcal{F}_R$ (disjoint union).

Therefore, for any value of s, and according to the definition of Hausdorff's measure:

$$\mathcal{H}^s(\mathcal{F}) = \mathcal{H}^s(\mathcal{F}_L) + \mathcal{H}^s(\mathcal{F}_R) = \left(\frac{1}{3}\right)^s \mathcal{H}^s(\mathcal{F}) + \left(\frac{1}{3}\right)^s \mathcal{H}^s(\mathcal{F}) \tag{1.25}$$

because $\mathcal{H}(\mathcal{F}_L) = \mathcal{H}(\mathcal{F}_R) = \frac{1}{3}\mathcal{H}(\mathcal{F})$ and $\mathcal{H}^s(\lambda \mathcal{F}) = \lambda^s \mathcal{H}^s(\mathcal{F})$ (Falconer, 1990).

As $s = \dim_H (\mathcal{F})$, $0 < \mathcal{H}^s(\lambda \mathcal{F}) < \infty$ (see Sections 1.1 and 1.2), it is possible to divide both sides of equation (1.25) by $\mathcal{H}^s(\mathcal{F})$ (different from 0). This yields:

$$1 = 2\left(\frac{1}{3}\right)^s \tag{1.26}$$

Hence: $s = \log.2 / \log.3 = 0.6309...$ (1.27)

An identical result can found by using the simplified definition:

$$s = \lim_{\delta \to 0} \frac{\log. M_\delta(\mathcal{F})}{\log. \frac{1}{\delta}} \tag{1.28}$$

and covering $\mathcal{F}$ with segments of length δ tending toward 0.

For the first iteration, two boxes of length 1/3 are sufficient to cover ε_1; $M_\delta(\mathcal{F}) = 2$ and $1/\delta = 3$, which yields:

$$D = \frac{\log.2}{\log.3} \tag{1.29}$$

Ch. 1]   Fractal Sets   21

or, for the m-th iteration;

$$D = \lim_{m \to \infty} \frac{\log. 2^m}{\log. 3^m} = \frac{\log.2}{\log.3} \qquad (1.30)$$

Some of the possible types of partition of [0 ; 1] are represented in Figure 1.5, defining sets with dimensions $0 < D < 1$.

1.3.4 Some Characteristic Properties of the Triadic Cantor Set

1. $\mathcal{F}$ is self-similar.
It is evident that the two intervals making the partition ε_1 of $\mathcal{F}([0 ; 1/3]$ and $[2/3 ; 1])$ are geometrically similar to $\mathcal{F}$, with a scale ratio of 1/3. Going further, the parts of $\mathcal{F}$ in each of the 4 intervals of ε_2 are similar to $\mathcal{F}$, with a scale ratio of 1/9, and so on. The Cantor Set contains copies of itself, at different scales.

2. $\mathcal{F}$ has a fine structure, which shows details at any arbitrarily small scale.
The larger the representation of a Cantor set, the more apparent the voids will become.

3. Although $\mathcal{F}$ has a complicated structure, its actual definition is very precise.

4. $\mathcal{F}$ is formed by a recursive process. The successive steps produce approximations ε_k of $\mathcal{F}$ which become progressively better as k increases.

5. $\mathcal{F}$'s geometry cannot be easily described with traditional terms: $\mathcal{F}$ is no geometric assembly of points satisfying a simple geometric constraint, nor is it the solution of a simple equation.

6. The local geometry of $\mathcal{F}$ is difficult to describe. Near each of its points, there is a high number of other points separated by empty spaces of variable lengths.

7. Although $\mathcal{F}$ can be described as a very large set (it is infinite and uncountable), its size cannot be quantified by measures such as the length. $\mathcal{F}$ has a null length[2] when the number of iterations tends towards ∞.

1.3.5 Random Cantor Set

The triadic Cantor set can be randomised in several ways. For example, each interval is divided into 3 equal parts, and the element to be removed is randomly selected; it may

---

[2] For example, if we take the interval [0 ; 1] = 1000 km. After 10 iterations, the cumulated length of the 1024 segments, each 0.0169 km long, is 17.3 km. After 20 iterations, 1,048,576 segments of length $2.869 \ 10^{-10}$ km have a cumulated length of 0.3 km. After 30 iterations, this cumulated length is only 5.21 m.

not be the median third, and there are three possibilities at each iteration (Figure 1.6). These different sets will still be characterised by a fractal dimension, which can for example be quantified with a box-counting technique (see Section 1.3.7).

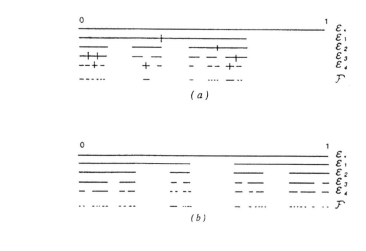

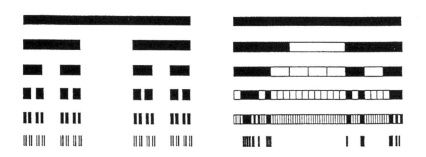

Figure 1.6. Random Cantor sets. Two ways to randomise a Cantor set are: (a) each interval is divided into 3 equal parts, the 2 intervals to remain are randomly selected; (b) each interval is replaced by 2 sub-intervals of random length. From Mandelbrot (1975), Dubois and Cheminée (1988), and Falconer (1990).

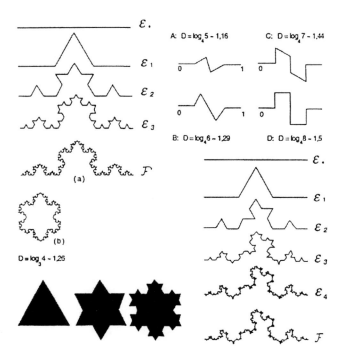

Figure 1.7. Von Koch curves. These curves can be obtained with several generators (A, B, C, D). The snowflake (B) and the randomised Von Koch curve (D) are shown at the bottom of the figure. From Mandelbrot (1975) and Falconer (1990).

1.3.6 Other Examples of Fractal Sets

Von Koch Curve

With a dimension greater than 1, this set helps understand the fractal character of a rocky coastline or a river path. The Von Koch curve is started from a segment ε_0 of length l_0. Its median third is replaced by the two sides of an equilateral triangle, whose third side was the median interval (Figure 1.7a). This makes up ε_1, with 4 segments. If the process is repeated on each segment of ε_1, we get ε_2, with 16 segments of identical lengths. Successive iterations produce $\varepsilon_3, \varepsilon_4, ..., \varepsilon_n$. The resulting curve becomes highly jagged (Figure 1.7a). Performing the same operations on the 3 sides of an equilateral triangle would produce a *snowflake* figure (Figure 1.7b). The fractal dimension of the Von Koch curve is log.4/log.3 = 1.262 (for 4 segments of length 1/3).

As for Cantor's set, an infinity of Von Koch curves can be generated by varying the number of segments and their emplacement (see examples in the right half of Figure 1.7), but the fractal dimension will always be 1 < D < 2. It is also possible to randomise this construction process, by randomly determining at each iteration on which segment side to operate (Figure 1.7d).

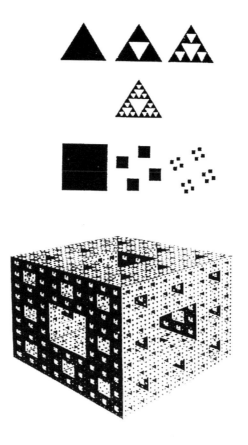

Figure 1.8. Classic fractal sets. Some of the best known fractal sets are Sierpinski's Triangle (top), the Cantor Dust in the plane (middle), and in space (bottom), also called Menger Sponge. These sets have respective fractal dimensions of 1.585, 1, and 2.72.

Sierpinski's Triangle

Inside an equilateral triangle, it is possible to inscribe another equilateral triangle half smaller. This smaller surface can be removed, and 3 half-smaller triangles inscribed and removed in the 3 remaining filled triangles (Figure 1.8). The iterative process creates a structure called Sierpinski's Triangle (see also Section 4.1). The fractal dimension of this construction is $1 < D < 2$; it is equal to $\log 3 / \log 2 = 1.5851$.

Other geometric figures can form fractal sets. Sierpinski's Carpet is the extension to the plane of the triadic set, and the Menger Sponge is its extension to space (Figure 1.8). Their respective fractal dimensions are D = log.2 / log.3 for the triadic Cantor set, D = log.8 / log.3 for Sierpinski's Carpet, D = log. 20 / log.3 for the Menger Sponge.

1.3.7 Computation of the Fractal Dimension

In the simple cases where the fractal set $\mathcal{F}$ results from a hierarchical geometric process, the fractal dimension D can be computed simply by refering to the relation $N = C / r^D$, or by using the Hausdorff measure (see Section 1.3).

But "box-counting" is the method generally used. For example, in the case of a randomised Cantor set, $\mathcal{F}$'s domain can be covered by successive segments of length r_i, and one counts either the number of segments N_i containing "dust", or the ratio x_i of these segments N_i to the total number of segments covering the interval occupied by $\mathcal{F}$ (see in Section 2.5 the study of time series).

If $\mathcal{F}$ is a non-empty subset of $\mathcal{R}^n$, let $N_\delta(\mathcal{F})$ be the smallest number of elements (boxes, or spheres, or others) of diameter δ which can cover $\mathcal{F}$. The smallest and largest box-counting dimensions are:

$$\underline{\dim}_B(\mathcal{F}) = \lim_{\delta \to 0} \frac{\log.(N_\delta(\mathcal{F}))}{-\log.\delta} \qquad (1.31)$$

$$\overline{\dim}_B(\mathcal{F}) = \overline{\lim}_{\delta \to 0} \frac{\log.(N_\delta(\mathcal{F}))}{-\log.\delta} \qquad (1.32)$$

Falconer (1990) demonstrates that $\dim_B(\mathcal{F})$ falls between these two values and that its definition is therefore:

$$\dim_B(\mathcal{F}) = \lim_{\delta \to 0} \frac{\log.(N_\delta(\mathcal{F}))}{-\log.\delta} \qquad (1.33)$$

where $N_\delta(\mathcal{F})$ can be any of the following:
- the smallest number of closed spheres of radius δ which cover $\mathcal{F}$
- the smallest numbers of cubes of length δ which cover $\mathcal{F}$
- the number of cubes of the grid of step δ which intersects $\mathcal{F}$
- the smallest number of elements of diameter $\leq \delta$ which cover $\mathcal{F}$
- the largest number of disjoint spheres of radius δ which are centred in $\mathcal{F}$

This list could be extended to all objects best suited to the "covering" of $\mathcal{F}$. And the terms of spheres and cubes are of course meant in an Euclidean space with n dimensions, n maybe greater than 3.

Figure 1.9 shows these different coverings of $\mathcal{F}$. An example is also given in section 2.2 (Tectonics) for the study with two different techniques of the fractal dimension of San Andreas' faults, with n = 2.

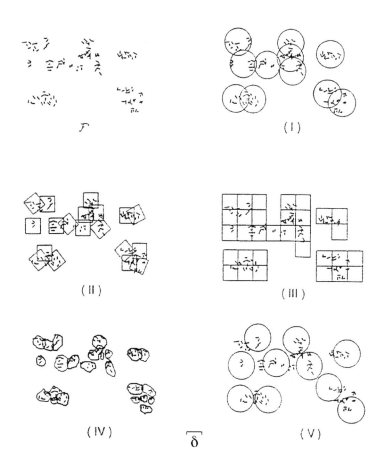

Figure 1.9. Computation of the fractal dimension by box-counting. Five different methods for covering and box-counting to compute the dimension of a set $\mathcal{F}$ were proposed and demonstrated by Falconer (1990).

1.3.8 A Few Properties of Fractal Sets: Projection, Product, Intersection

Projection of fractal sets

It is possible to describe intuitively Figure 1.10 (from Falconer, 1990), which presents the shadows of geometric objects in $\mathcal{R}^3$, i.e. their projection on a plane. A continuous linear curve, of dimension 1, has for shadow a line of dimension 1. Generally speaking, a surface (dimension 2) or a volume (dimension 3) have a shadow of dimension 2. It is therefore possible to think that a fractal set $\mathcal{F}$ in $\mathcal{R}^3$ has a projection of dimension 2 if $\dim_H(\mathcal{F}) > 2$ and of dimension $\dim_H(\mathcal{F})$ if $\dim_H(\mathcal{F}) < 2$ ($\dim_H$ being the Hausdorff dimension).

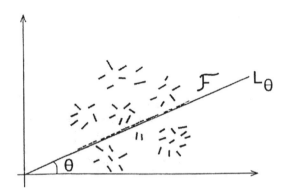

Figure 1.10. Projection of a fractal set. Here, a set $\mathcal{F}$ of $\mathcal{R}^2$ is projected on the line L_θ. The dimension of the projection is the dimension of $\mathcal{F}$ if the latter is less than 1, and 1 if the dimension of $\mathcal{F}$ is greater than 1. From Falconer, 1990.

Projection in $\mathcal{R}^2$

Let L_θ be a line passing through the origin of $\mathcal{R}^2$ and making an angle θ with the horizontal axis. Pr_θ is the projection on L_θ. If $\mathcal{F}$ is a subset of $\mathcal{R}^2$, $Pr_\theta(\mathcal{F})$ is the orthogonal projection of $\mathcal{F}$ on L_θ (Figure 1.10). We have:

$$|Pr_\theta(x) - Pr_\theta(y)| \leq |x-y|, \text{ if } x,y \in \mathcal{R}^2 \qquad (1.34)$$

Pr_θ verifies the Lipschitz condition: $|f(x) - f(y)| \leq c |x-y|^\alpha$, if $\alpha = 1$ (see Annex A for more details) and therefore:

$$\dim_H(Pr_\theta(\mathcal{F})) \leq \min.\left\{\dim_H(\mathcal{F}); 1\right\} \qquad (1.35)$$

This relation means that the Hausdorff dimension of the projection of the set $\mathcal{F}$ on the line L_θ is smaller than or equal to the smallest of the two: $\mathcal{F}$'s dimension, or 1. If $\mathcal{F}$ has a Hausdorff dimension greater than 1, the Hausdorff dimension of its projection on L_θ will equal 1. If $\mathcal{F}$ has a dimension smaller than 1, the dimension of its projection on L_θ will be the dimension of $\mathcal{F}$.

This is what the following theorem is expressing (Falconer, 1990):

> For a subset $\mathcal{F}$ of $\mathcal{R}^2$, $\mathcal{F}$ being a Borelian;
> 1. if $\dim_H(\mathcal{F}) \leq 1$, then $\dim_H(Pr_\theta(\mathcal{F})) = \dim_H(\mathcal{F})$, for any $\theta \in [0, \pi]$
> 2. if $\dim_H(\mathcal{F}) > 1$, then $\dim_H(Pr_\theta(\mathcal{F})) = 1$, for any $\theta \in [0, \pi]$

The importance of this theorem will be made clearer in Section 2.2.

<u>Projection in $\mathcal{R}^n$</u>

The generalisation in $\mathcal{R}^n$ considers $\mathcal{G}_{n,k}$, the set of k-dimensional subspaces (i.e. the "planes" of dimension k passing through the origin of $\mathcal{R}^n$). These subspaces are parameterised by k.(n-k) coordinates. The orthogonal projection on the plane Π of dimension k is noted Pr_Π. Note that, for the lack of another term, we keep the word of "plane" (which in Euclidean geometry has a dimension k = 2) for subspaces whose dimension may be greater than 2. By generalising the previous theorem, it is possible to demonstrate the following theorem:

> For a subset $\mathcal{F}$ of $\mathcal{R}^n$, $\mathcal{F}$ being a Borelian;
> 1. if $\dim_H(\mathcal{F}) \leq k$, then $\dim_H(Pr_\Pi(\mathcal{F})) = \dim_H(\mathcal{F})$, for any $\Pi \in \mathcal{G}_{n,k}$
> 2. if $\dim_H(\mathcal{F}) > k$, then $\dim_H(Pr_\Pi(\mathcal{F})) = k$, for any $\Pi \in \mathcal{G}_{n,k}$

This theorem is clearly the extension of the preceding one to n dimensions. If we take, for example, a subset $\mathcal{F}$ of $\mathcal{R}^3$ (see Figure 1.10) the plane projections of $\mathcal{F}$ are generally of dimension $\min.\{2, \dim_H(\mathcal{F})\}$ which introduces important practical applications. Now, it is indeed possible to estimate the dimension of an object in space by estimating its dimension on a photograph taken from any direction. Provided it is less than 2, it is possible to deduce the dimension of the object (by adding 1). Such a reduction makes it possible to search for the dimension of an object in space, as box-counting is much more difficult in 3 dimensions than in 2. Applications of this property will be demonstrated in tectonic studies (see Section 2.2), where the dimension of a fault plane will be computed by box-counting on the surface expression of the fault. The application of these theorems will be shown to avoid erroneous interpretations of, for example, fracture fields in rocks.

<u>Product of fractals</u>

If $\mathcal{E}$ is a subset of $\mathcal{R}^n$ and $\mathcal{F}$ a subset of $\mathcal{R}^m$, the product $\mathcal{E} \times \mathcal{F}$ is defined as the set of points whose first coordinate belongs to $\mathcal{E}$ and whose second coordinate belongs to $\mathcal{F}$:

$$\mathcal{E} \times \mathcal{F} = \left\{ (x,y) \in \mathcal{R}^{n+m}; \, x \in \mathcal{R}^n; \, y \in \mathcal{R}^m \right\} \quad (1.36)$$

Generally:

$$\dim_H(\mathcal{E} \times \mathcal{F}) = \dim_H(\mathcal{E}) + \dim_H(\mathcal{F}) \quad (1.37)$$

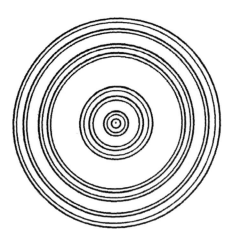

Figure 1.11. Cantor's Target. Its dimension is the sum of the dimensions of the triadic set and the circle, i.e. 1 + 0.6309.

If $\mathcal{E}$ is a unit interval in $\mathcal{R}^1$ and $\mathcal{F}$ a unit interval in $\mathcal{R}^1$, then the product $\mathcal{E} \times \mathcal{F}$ is a unit square in $\mathcal{R}^2$.

If $\mathcal{F}$ is a triadic Cantor set, the product $\mathcal{F} \times \mathcal{F}$ in $\mathcal{R}^2$ verifies :

$$\dim_H(\mathcal{F} \times \mathcal{F}) = 2 \dim_H(\mathcal{F}) = 2 \log.2 / \log.3 \quad (1.38)$$

Cantor's Target is obtained by making the product of $\mathcal{E}$, a triadic Cantor set for which $\dim_H(\mathcal{E}) = \log.2 / \log.3$, and of $\mathcal{F}$, a circle of dimension 1. This constructs a target in $\mathcal{R}^2$, with a dimension $1 < D < 2$ (Figure 1.11).

$$\mathcal{T} = \mathcal{E} \times \mathcal{F} \Longrightarrow \dim_H(\mathcal{T}) = \dim_H(\mathcal{E}) + \dim_H(\mathcal{F}) = \frac{\log.2}{\log.3} + 1 \quad (1.39)$$

<u>Intersection of fractals</u>

The intersection of two fractal sets is often a fractal, whose dimension can be computed.

If $\mathcal{F}_1$ is a copy of $\mathcal{F}$, two cases are possible: $\dim_H(\mathcal{F} \cap \mathcal{F}_1) = \dim_H(\mathcal{F})$ if $\mathcal{F}_1 = \mathcal{F}$, or $\dim_H(\mathcal{F} \cap \mathcal{F}_1) = 0$ if $\mathcal{F}$ and $\mathcal{F}_1$ are disjoint.

One example is the unit segment $\mathcal{F}$. If $\mathcal{F}_1$ is an acceptable copy of $\mathcal{F}$, $\mathcal{F} \cap \mathcal{F}_1$ may be a segment only in the rare case where $\mathcal{F}$ and $\mathcal{F}_1$ are colinear. If $\mathcal{F}$ and $\mathcal{F}_1$ are crossing at a certain angle, $\mathcal{F} \cap \mathcal{F}_1$ is a single point, but $\mathcal{F} \cap \mathcal{F}_2$ is also a point for all acceptable copies $\mathcal{F}_2$ of $\mathcal{F}$ close enough to $\mathcal{F}_1$. The most frequent situation is where $\mathcal{F} \cap \mathcal{F}_1$ is restricted to one point.

In spaces with more dimensions, the results are similar. In $\mathcal{R}^3$, the intersection of two surfaces is a curve, and the intersection of a curve and a surface is a point. The intersection of two curves is generally empty.

In $\mathcal{R}^n$, if the multi-dimensional sets $\mathcal{E}$ and $\mathcal{F}$ have an intersection, it generally is a subset of dimension $\max.\{0, \dim.\mathcal{E} + \dim.\mathcal{F} - n\}$. More precisely, if $\dim.\mathcal{E} + \dim.\mathcal{F} - n > 0$, then $\dim.(\mathcal{E} \cap \sigma(\mathcal{F})) = \dim.\mathcal{E} + \dim.\mathcal{F} - n$ for rigid transformations σ, and $\dim.(\mathcal{E} \cap \sigma(\mathcal{F})) = 0$ for any other σ.

Two theorems formalise these properties of fractals' intersections. They were demonstrated by Falconer (1990):

> Theorem 1: Let $\mathcal{F}+x$ be the translation of $\mathcal{F}$ by a vector x. If $\mathcal{E}$ and $\mathcal{F}$ are Borelian subsets of $\mathcal{R}^n$, then:
>
> $$\dim_H(\mathcal{E} \cap (\mathcal{F}+x)) \leq \max.\left\{0;\ \dim_H(\mathcal{E} \times \mathcal{F}) - n\right\} \quad (1.40)$$
>
> for nearly all $x \in \mathcal{R}^n$.

> Theorem 2: Let $\mathcal{E}$, $\mathcal{F}$ be two Borelian subsets of $\mathcal{R}^n$, and let $\mathcal{G}$ be a group of transformations on $\mathcal{R}^n$. For a set of movements $\sigma \in \mathcal{G}$ (of positive measure), and in the following cases:
>   1. $\mathcal{G}$ is the group of similarities, and $\mathcal{E}$ and $\mathcal{F}$ are arbitrary sets.
>   2. $\mathcal{G}$ is the group of rigid movements, $\mathcal{E}$ is arbitrary and $\mathcal{F}$ a curve, or a surface, or a set with a higher dimension.
>   3. $\mathcal{G}$ is a group of rigid movements, $\mathcal{E}$ and $\mathcal{F}$ are arbitrary, and either $\dim_H(\mathcal{E}) > (n+1)/2$ or $\dim_H(\mathcal{F}) > (n+1)/2$
>
> Then: $\dim_H(\mathcal{E} \cap \sigma(\mathcal{F})) \geq \dim.\mathcal{E} + \dim.\mathcal{F} - n$

It is easy to guess all the interest brought by these geometrical properties: projection, product, intersection of fractal sets, for the studies of rock fracturation and tectonics. Indeed, as will be seen in section 2.2, fault fields are fractal sets in $\mathcal{R}^3$, but their study is rarely performed in this 3-D space, and rather with 2-D surveys and photographs, or even 1-D drillings. For these reasons, the developments of the geometrical approaches to fractal structures have been added to the Annexes. They are extracted from the seminal work of Falconer (1985, 1988, 1990). The reader might also look at Besicovitch (1939), Marstrand (1954), Kaufman (1968) and Matila (1975, 1984, 1985).

1.4 DYNAMIC SYSTEMS

> *Time sparkles, and the Dream becomes knowledge*
> Paul Valéry, Le Cimetière Marin

The domain of dynamic systems is very large and has given rise to numerous studies during recent years. Dynamic systems are now widely used to model biological, geographic or economic processes, along with their traditional domains of engineering and physics. The present section will start with simple examples of mechanical systems whose behaviour can become chaotic, and more general definitions will be given later. We follow here the path of Bergé et al. (1984) and Croquette (1987), who showed that most mechanical systems fulfil the conditions that may lead to chaos. The harmonic oscillator, the simple pendulum, the coupled harmonic oscillators and pendulum systems with central and non-central force fields will be successively described.

1.4.1 The Harmonic Oscillator

The mass M moves along the Oz axis and is recalled toward the origin by a spring of tautness K (Figure 1.12). In the absence of friction, the movement of the mass will be

described by:

$$M\frac{d^2z}{dt^2} = K\,z \Rightarrow z = z_0 \cos(\omega_0 t + \Phi) \tag{1.41}$$

where z_0 is the movement's amplitude, $\omega_0 = \sqrt{K/M}$ its pulsation, and Φ an arbitrary phase. For the solution of the second-order differential equation to be unique, the two initial conditions z_0 and Φ need to be fixed. As the mass can only move along-axis, and only one variable is necessary to describe its position, this system is said to have *one degree of freedom*, and is governed by a linear equation.

The phase space is formed with one coordinate being the variable describing the movement and the other coordinate its time derivative (see Annex A.3.1 for detailed definitions). More precisely, in this case, the first coordinate axis corresponds to z(t), position of the moving mass, and the second coordinate axis corresponds to the impulsion of the mass Mdz(t)/dt.

This two-dimensional space is sufficient to represent graphically the movements of the oscillator. At any time, the state of the system can be fully described. The succession with time of the different states {z, Mdz/dt} forms a *phase trajectory* in the phase space. The position varies as $\cos(\omega_0 t + \Phi)$ and the impulse as $\sin(\omega_0 t + \Phi)$: they are in quadrature and the phase trajectories are ellipses (Figure 1.12).

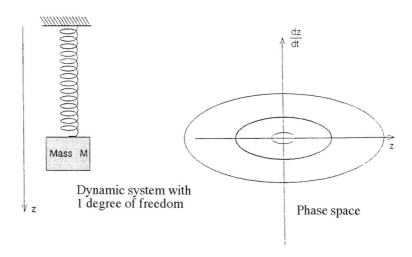

Figure 1.12. The harmonic oscillator. It is composed of a mass M and a spring, and has only one degree of freedom (M can only move along the Oz axis). The graph on the right shows the phase space and the phase trajectories.

1.4.2 The Simple Pendulum

A mass M, at the end of a rod of length l, can turn freely and without friction around an axis (Figure 1.13). Only one variable is necessary to describe the position of the mass: the angle θ between the rod and the vertical. This system is *non-linear*, as the return varies with sin.θ, not with θ.

The movement's differential equation is:

$$M\frac{d^2\theta}{dt^2} = \frac{g}{l}\sin\theta \tag{1.42}$$

where g is the acceleration of gravity.

The phase trajectory belongs to a two-dimensional space (θ(t); dθ(t)/dt). The pendulum's energy can be expressed as Ml [l (dθ/dt)² + g cosθ]. The energy of the system is invariant along phase trajectories, as had already been remarked by Croquette (1987). Figure 1.13 shows a few of these trajectories, which are contour lines of the system's energy.

The pendulum's non-linear character is revealed by the dependency of the periods of oscillation on the initial conditions. They increase with amplitude, and the phase trajectories are deformed versions of the ellipses from the linear case (Figure 1.13). In this graph, the trajectories inside the domain bounded by the two limit trajectories correspond to complete rotations of the pendulum around its rotation axis.

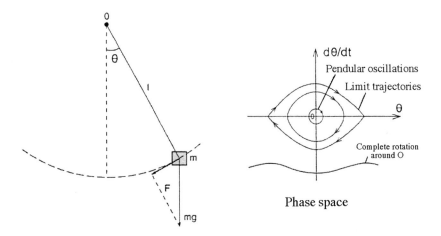

Figure 1.13. The simple pendulum, its phase space and the phase trajectories. Limit trajectories mark the boundary beyond which the trajectories become full rotations around the rotation axis.

1.4.3 Coupled Harmonic Oscillators

This system is a variation on the first example shown, with the coupling of two oscillators with a small spring of tautness k joining the two masses. This is a system with *two degrees of freedom*, as two variables Z_1 and Z_2 are necessary to define the system's position. This system is *linear*, and is governed by the two differential equations:

$$\begin{cases} M\dfrac{d^2Z_1}{dt^2} = K\,Z_1 + k\,(Z_1 - Z_2) \\ \\ M\dfrac{d^2Z_2}{dt^2} = K\,Z_2 + k\,(Z_2 - Z_1) \end{cases} \quad (1.43)$$

By substituting the variables $X = Z_1 + Z_2$ and $Y = Z_1 - Z_2$, we get a system with two independent differential equations:

$$\begin{cases} M\dfrac{d^2X}{dt^2} = K\,X & (1.44) \\ \\ M\dfrac{d^2Y}{dt^2} = (K + 2k)\,Y & (1.45) \end{cases}$$

Because the system is linear, it was possible to find two variables that would split this system with two degrees of freedom into two systems with one degree of freedom, and pulsations of $\sqrt{K/M}$ for X and $\sqrt{(K+2k)/M}$ for Y.

1.4.4 Systems with Central and Non-Central Force Fields

Figures 1.14 and 1.15 show two magnetic pendulums. In both cases, the systems have *two degrees of freedom* as two variables are necessary to define the position of the body in movement. These systems are also *non-linear*, because magnetic interaction is dipolar and not linear.

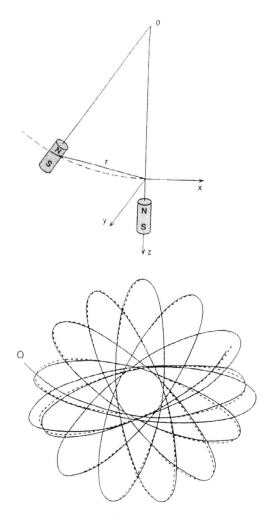

Figure 1.14. Magnetic pendulum with a central force field. Note the similarity between the trajectories, which shows a small sensitivity to the initial conditions. From Croquette, 1987.

The two position variables are r, distance to the vertical axis Oz, and θ, the angle between the vector **r** and a reference direction (Figures 1.14 and 1.15).

1. In the system submitted to a central force field, we have two invariants, the kinetic moment $I = m r^2 d\theta/dt$ and the total energy:

$$E = \frac{1}{2} m \left[\left(\frac{dr}{dt}\right)^2 + r \left(\frac{d\theta}{dt}\right)^2 \right] + V(r) \qquad (1.46)$$

where V(r) is the potential of the central force field. Introducing I in the expression of E, the variable θ can be eliminated and the result is a first-order differential equation in r:

$$\left(\frac{dr}{dt}\right)^2 + \frac{I^2}{2mr^3} + 2\left(\frac{V(r) - E}{m}\right) = 0 \qquad (1.47)$$

The actual physical trajectory is a projection of the pendulum's case in the phases' space. Its expression is given by Landau and Lifschitz (1969). In the case of a central force field, the invariants are the kinetic moment and the energy. When the force field is no longer central, the kinetic moment stops being invariant, and it is not possible to separate the system into two systems with one degree of freedom.

The first-order differential equation corresponds to an oscillator with one degree of freedom, submitted to a potential $V(r) + I^2 / 2mr^3$. This differential equation can be integrated, and the evolution of θ deduced from the expressions of I and r(t). Therefore, the existence of a constant kinetic moment makes it possible to separate this system with two degrees of freedom into two independent systems with one degree of freedom (as for the two coupled oscillators). The pendulum's movement is therefore made of two non-linear oscillations, with base frequencies f_1 and f_2 and their harmonics pf_1+qf_2 (p and q integers).

The phase space of the system has *a priori* four dimensions, with the axes r, dr/dt, θ, dθ/dt. Because the energy and kinetic moment are invariant, the space available for the phase trajectories decreases to two dimensions. The phase trajectory is therefore inscribed on a T^2 toroid corresponding to the superposition of the two oscillations.

2. In the case of the non-central force field, the kinetic moment is not invariant any more, and the system cannot be separated into two systems with one degree of freedom.

In this case, as in the previous case, the trajectory may be studied by a photographic process. Starting with identical initial conditions for two successive tests, one finds that in the case of a central force field, the two trajectories are very close and very slightly separate from each other, only because of the unavoidable infinitesimal differences in the initial conditions. However, in the case of the non-central force field, two close trajectories will diverge very rapidly and seem to evolve completely independently. This is experimentally bringing into light a basic property of dynamic systems: the *sensitivity to the initial conditions*.

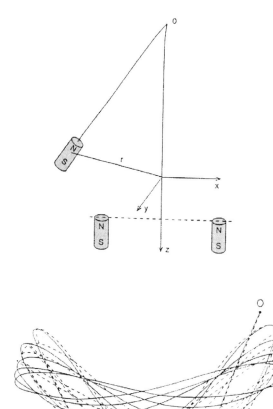

Figure 1.15. Magnetic pendulum with a non-central force field. Two trajectories starting from close initial conditions (point O) are rapidly diverging. From Croquette, 1987.

Two regular trajectories, with similar initial conditions, will diverge in the phase space, the distance between the two increasing approximately linearly with time, as is the case for the central force field. In the case of a chaotic trajectory, this distance will increase exponentially with time (the actual increase will be quantified later, during the study of Lyapunov's coefficients); this is why the two trajectories will rapidly become decorrelated.

Another way to describe these two cases consists in examining the trajectories in the phase space. In the first case (central force field), the trajectories are inscribed on a $\mathcal{T}^2$

toroid whereas in the second case (non-central force field), only the energy is invariant and the trajectory is included in a subspace of dimension 3. The chaotic trajectory differs from the regular trajectory by the Fourier spectra of the coordinates. For a central force field, the Fourier spectrum of each coordinate is formed of fine lines, at frequencies pf_1+qf_2 (p and q integers), whereas for a non-central force field, the Fourier spectrum is formed of a frequency continuum, similar to noise.

In all these examples, only the system with a non-central force field is chaotic, as its movement cannot be integrated. However, each time a system with N degrees of freedom can be separated into N independent systems with one degree of freedom, the movement can be integrated.

1.4.5 Summary

- Harmonic oscillator: 1 degree of freedom; the system is governed by a linear equation.

- Simple pendulum: 1 degree of freedom; non-linear system.

- Coupled harmonic oscillators: 2 degrees of freedom; linear system equivalent to 2 independent systems with 1 degree of freedom.

- Pendulum in a force field:

    - central: 2 degrees of freedom, linear system equivalent to 2 independent systems with 1 degree of freedom.

    - non-central: 2 degrees of freedom; non-linear system, *chaos*.

1.5 LIMIT CYCLES AND ATTRACTORS

> All appearance ...
> incites us to caution
> *Wittgenstein, 1937*

Other examples allow to introduce the new notions of limit cycles and attractors, refining at the same time our study of dynamic systems.

1.5.1 The Attraction Process

Let us consider a pendulum kept oscillating, like the one in a clock. Energy is provided to the pendulum (by way of an electromagnet, impulse, spring or weight mechanism) to compensate for the energy lost by friction in the system. When the energy provided equals the energy dissipated, a permanent state is established. As the movement is periodic, a limit cycle C is continuously described in the phase space by the point $(\theta(t), d\theta/dt)$.

If the pendulum is moved away from its permanent position, for example with a shock giving him an amplitude θ_1 and a speed $d\theta_1/dt$ greater than the values θ_m and $d\theta_m/dt$ of the limit cycle, the dissipation of energy with time will be stronger. Because of the higher amplitude, this energy will be greater than the constant energy provided to the system. Therefore, the phase trajectory will converge towards the permanent trajectory, along which the provided and dissipated energies are equal.

Conversely, if the pendulum is briefly slowed down, the amplitude θ_2 and speed $d\theta_2/dt$ will be smaller than θ_m and $d\theta_m/dt$. The energy provided will be greater than the energy dissipated in shorter oscillations. And the phase trajectory will tend rapidly back toward the curve C (Figure 1.16).

Any trajectory starting from a point $(\theta_i ; d\theta_i/dt)$ different from $(\theta_m ; d\theta_m/dt)$ will be attracted toward the asymptotic trajectory. Hence the term of *limit cycle* given to C, to describe the attraction exerted by this object. For practical reasons, θ_i and $d\theta_i/dt$ cannot be given any value. For attraction to happen, these values need to belong to what is called an *attraction basin*.

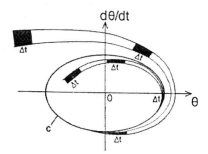

Figure 1.16. Area contraction. For a pendulum kept oscillating, the two trajectories converge toward the attracting limit cycle. The areas corresponding to the same Δt are shrinking. There is loss of information.

1.5.2 Area Contraction

Using the previous example, and reasoning in the phases space $(\theta ; d\theta/dt)$, we can start from two initial positions, close but still different, located for example outside the limit cycle (Figure 1.16). During the time interval Δt, an area A_1 can be defined between two neighbouring trajectories. Later, this area will be reduced to a value A_2 for the same Δt, because of the convergence toward the limit cycle C. Even later, this area will be further reduced to a value A_3: $A_1 > A_2 > A_3$. There is *area contraction*. The areas are decreasing until they correspond to a segment at the arrival on the attractor.

The first consequence is the loss of information about the relative position of the points (Figure 1.16). When the attractor is encountered, this information disappears. The exact position of the initial point is forgotten.

At the beginning, two coordinates θ and $d\theta/dt$ were necessary to characterise the system. Its phase space had two dimensions. Once on the attractor, or limit cycle, there is only one possible trajectory. A single curvilinear coordinate is enough to locate the point along the curve. This is a general property of attractors: the dimension d of the attractor is smaller than the dimension n of the phases' space, i.e. the number of degrees of freedom of the dynamic system: $d < n$.

1.5.3 Aperiodic Attractors, Chaotic Regime, Strange Attractors

During the study of mechanical systems, we saw that the coordinates of an oscillator in a central force field displayed a Fourier spectrum with a frequency continuum, like a noise. This shows the erratic, disorderly character of the movement. The disorder rate can be estimated by introducing a function which measures the similarity of the variable X, at a time t, with its value at a time $t+\tau$:

$$C(\tau) = \frac{1}{t_2 - t_1} \int_{t_1}^{t_2} X(t).X(t+\tau) dt \qquad (1.48)$$

i.e. :

$$C(\tau) = < X(t).X(t+\tau) > \qquad (1.49)$$

The function $C(\tau)$ can be built with the successive values of τ; it is a temporal autocorrelation function. It was shown by Wiener and by Kinchine that $C(\tau)$ is the Fourier transform of the power spectrum. If X(t) is constant, periodic or quasi-periodic, then $C(\tau)$ will be different from zero when τ tends towards $+\infty$. In this case, the system's behaviour is predictable.

Conversely, for an aperiodic chaotic system, the power spectrum will display a continuous interval, and $C(\tau)$ will tend toward zero when τ increases. The similarity of the signal with itself will diminish with time, and disappear for times long enough. A result is that the knowledge of X(t), during a period of time as long as one wants, does not allow one to predict the future behaviour of a later X(t). The system is chaotic and unpredictable because of its progressive loss of internal similarity.

It will be shown later that the degree of "unpredictability" is not that absolute. For the Lorenz attractor (a determinist chaotic system), Keilis-Borok (1990) demonstrated that the jump from one lobe to the other, of a point in the phase trajectory on the attractor was subject to some repeatability in a relatively narrow part of the same lobe. This could therefore enable a prediction of some sort. Recent studies have made use of these properties with the notion of intermittence (Sujihara and May, 1990; Sornette et al., 1991).

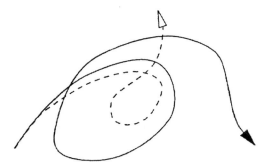

Figure 1.17. The sensitivity to initial conditions. The divergence of two initially close phase trajectories leads unpredictably to a multitude of final states; there is a gain of information (also see the example in Section 4.2.1)

With the oscillator in a near-central force field, we saw the loss of information about the initial conditions. Although starting from several initial states, away from the attractor, the phase trajectories were finally always merging with the attractor. Once this has been accomplished, the attractor can, however, lead to a multitude of final states: there is a gain in information. Once consequence is that two trajectories on the attractor, at first extremely close, will necessarily diverge and lose all similarity after some time (Figure 1.17).

If a system is represented by an attractor on which trajectories diverge, then this system is chaotic. This amplification[3] of errors or uncertainties about the initial conditions in a chaotic system s named the *Sensitivity to the Initial Conditions (SIC)*. The notion of an attractor with SIC presents a double paradox:

- The first paradox resides in the antinomy between attraction, which implies a convergence of trajectories, and the SIC which implies their divergence. There must be a minimal dimension for the attractor, i.e. a minimal dimension for the phase space. The information-creating divergence acts as some sort of balance to counteract the information-losing convergence.

- The second paradox is that, for the SIC to happen, the attractor's dimension must satisfy $d > 2$. But, in a dissipative system, there is area contraction (cf. the pendulum) or volume contraction (if $d > 2$) in the phases' space. The attractor's volume must therefore be zero, which means that in a three-dimensional space its dimension must verify $d > 3$. Therefore, an attractor susceptible of being chaotic with SIC must have a dimension $2 < d < 3$. This is not possible in an Euclidean space, but may exist in a space with a non-integer or fractal dimension.

---

[3] exponential. See Section 1.5.6 about Lyapunov's coefficients.

Such attractors were introduced by Ruelle and Takens (1971), and are called *Strange Attractors*.

In summary, a dynamic system may become chaotic if the dimension of the phase space is greater or equal to 3. Such a chaos, with a small number of degrees of freedom, stems from the SIC of phase trajectories along attractors with three properties:

- Phase trajectories converge *toward* the attractor.

- Neighbouring phase trajectories diverge *on* the attractor (SIC)

- The attractor's dimension in the phase space is fractal.

1.5.4 Practical Search for an Attractor and its Dimension

Experimental observations show that many dynamic systems reach a chaotic state after a small number of periodic or quasi-periodic oscillations of a variable characterising the system's evolution with time. The main problem is to determine if this chaos comes from a very large number of degrees of freedom, and is therefore random, or if it is determinist, i.e. characterised by a small number of degrees of freedom. If such is the case, it can be expected that the variable's phase trajectories have an attractor with a fractal structure and the properties of strange attractors described above (Malraison et al., 1983). Bergé et al. (1984) show that, starting from a temporal variation $X(t)$, it is possible to build a trajectory in a p-dimensional space with the coordinates $X(t)$, $X(t+\tau)$, $X(t+2\tau)$, ..., $X(t+(p-1)\tau)$ where τ is a time delay suitably chosen.

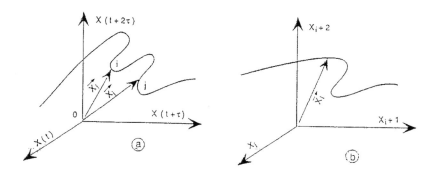

Figure 1.18. Phase trajectory in a three-dimensional space. (a) each point has $X(t)$, $X(t+\tau)$, $X(t+2\tau)$ for coordinates; (b) in this graph, time is replaced by the occurrence numbers of the events.

In some cases where the signal cannot be digitised, we have a succession of values X_i, e.g. the time between two events (earthquakes, volcanic eruptions, geomagnetic inversions). A pseudo-phase space is built with p dimensions by taking for each point the coordinates: $X_i, X_{i+1}, ..., X_{i+p-1}$, i being the number of the event, (i+1) the number of the next event, etc. The delay τ is replaced by the occurrence number of the event (Figure 1.18).

For a chaotic system, the positions along the same trajectory of two points, far from each other in time, are completely uncorrelated. However, if all points are located on an attractor, there is between them a spatial correlation which can be quantified with an appropriate correlation function. Grassberger and Procaccia (1983) suggest looking at the asymptotical behaviour in r^v of the integral correlation function:

$$C(r) = \lim_{m \to \infty} \frac{1}{m^2} \int_{i;j=1}^{\infty} H(r) - |X_i - X_j| \tag{1.50}$$

where H is the Heaviside function, and i and j are the two indices of the points along the m-point trajectory.

$$C(r) = \lim_{m \to \infty} \frac{1}{m^2} \{ \text{number of pairs (i;j) for which } |X_i - X_j| < r \} \tag{1.51}$$

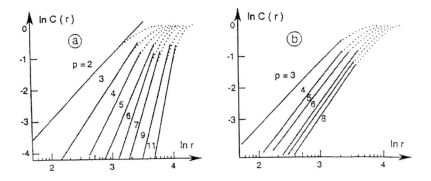

Figure 1.19. The attractor's dimension. The two graphs log.C(r) vs log(r) are drawn for increasing dimensions p of the phase space, in the case of a white noise and in the case of a system with an attractor. From Malraison et al. (1983), and Dubois and Cheminée (1992).

If $\log.C(r) = v \log(r)$, it means that $C(r) = r^v$. We first build the phase trajectories for integer values of the phases' space dimension p, successively 1, 2, 3, 4, 5, ..., p. For each value of p, we compute C(r) from the distances:

$$|X_i - X_j| = \left[(X_i-X_j)^2+(X_{i+1}-X_{j+1})^2+...+(X_{i+(p-1)}-X_{j+(p-1)})^2\right] \quad (1.52)$$

And, last, we look at the slope of the curve: $\log.C(r) = f(\log(r))$ (Figure 1.19).

For a white noise, or for a dynamic system with many degrees of freedom, the exponent ν will continue to increase as p increases. Conversely, if ν becomes independent from p, then the system is chaotic; the chaos is determinist and the corresponding attractor is a strange attractor of dimension ν. Plotting ν as a function of p shows there is a value of p after which ν stays at a constant level; there is "saturation" for the phase spaces of dimension $\geq p$.

Atten and Caputo (1987) demonstrated there is a relation between ν and its saturation dimension: $p_s \geq 2\nu + 1$ (Figure 1.20). For example, in the case of Hénon's attractor (dimension $\nu = 1.26$), the saturation dimension of the phase space will be $p_s \approx 4$. It will therefore be necessary to carry computations beyond this value of p.

The number of points necessary to compute the attractor's dimension must be high, several thousands or tens of thousands. But this is not always feasible, for example with the short series of recorded volcanic eruptions or strong earthquakes in a given region. This also proves to be a problem for dynamic systems in which the uncertainty on results obtained on long time series may be due to the non-stationarity of the system. In this case, it is necessary to use short series (for example in a moving window) on short time intervals in which the system may be considered stationary.

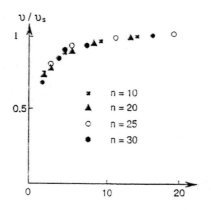

Figure 1.20. Relation of Atten and Caputo (1987) in a grid turbulence case. The values of ν were normalised by the saturation value ν_s; $p = 2\nu_s + 1$.

Many studies have been devoted to this problem (Guckheimer, 1986; Havstad and Ehlers, 1989; Möller et al., 1989; Liebovitch and Toth, 1989; Ramsey and Yuan, 1990). In a study of the theoretical system of MacKey-Glass (1977) with an attractor of well-established dimension 6.7, Havstad and Ehlers (1989) demonstrated that in specific conditions it was possible to go down to a series of 50 vectors, the computed dimension being less than 5% apart from its actual value ... Dubois and Pambrun (1990) show with Hénon's attractor that, after 100 vectors, the value found for v is as close as 5% to the actual value $v = 1.26$.

Note: There are other methods of computing an attractor's dimension. One of these methods consists in computing the Lyapunov coefficients and using the relations of Kaplan and Yorke (1978) and Farmer et al. (1983) (see Section 1.5.6).

1.5.5 Examples of Strange Attractors

So far, we have seen periodic attractors and only mentioned aperiodic attractors with a chaotic behaviour. The first of the next examples is at the origin of the definition.

<u>Lorenz Attractor</u>

After some simplifications, the model of Lorenz (1963) describes the dynamic behaviour of a convecting fluid with three differential equations, defining a three-dimensional flow (X,Y,Z) (the demonstration can be found in Bergé et al., 1984):

$$\begin{cases} \frac{dX}{dT} = P_r Y - P_r X \\ \frac{dY}{dT} = -XZ = rX - Y \\ \frac{dZ}{dT} = XY - bZ \end{cases} \quad (1.53)$$

in which P_r is the fluid's Prandtl number $P_r = v / D_T$, v is the fluid's kinematic viscosity, D_T its thermal diffusivity at temperature T, and $b = 4\pi^2 / (\pi^2 + q^2)$ where q is the pulsation in the direction of the movement of period $2\pi / q$. For liquids and water, P_r is between 5 and 10, depending on the temperature. For silicon oils, $P_r > 100$. To study the system, one generally takes values of $P_r = 10$ and $b = 8/3$, using r as a control parameter.

The system (1.53) is generally not integrable analytically. But it is solvable numerically, starting from initial values (X(0), Y(0), Z(0)) and building the flow (X(t), Y(t), Z(t)) step by step, along with the corresponding trajectories. These phase trajectories are always on the same geometrical object, made up of two lobes and well known in scientific journals (Figure 1.21). This object is a strange attractor corresponding to a chaotic system, and the trajectories jump from one lobe to the other after a few cycles in each.

The Lie derivative of a system is 1/V dV/dT, where V is the volume defined in the phase space by the phase trajectories. The variation of volume under the action of the flow is:

$$\frac{1}{V}\frac{dV}{dT} = \sum_{i=1}^{n} \frac{\partial \dot{X}_i}{\partial X_i} \quad (1.54)$$

In the case of the Lorenz model and its three-dimensional space (with an attractor of dimension 2.06, $P_r = 10$ and $b = 8/3$):

$$\frac{\partial \dot{X}}{\partial X} + \frac{\partial \dot{Y}}{\partial Y} + \frac{\partial \dot{Z}}{\partial Z} = -(P_r + b + 1) = -\frac{41}{3} \qquad (1.55)$$

After one time unit, the volume's contraction is $e^{-41/3} \approx 10^{-6}$. The contraction is very rapid, and the model is said to be *highly dissipative*.

The attractor's dimension is 2.06, larger than the Euclidean dimension of the surface. It cannot be represented in a plane, but can be represented very easily in a 3-D space where it does not fill an entire volume. We will see later (Chapter 2) that, for these reasons, it is possible to submit this attractor to a first-return map, which will allow its visualisation on a Poincaré section $Z = $ cst, in the plane (X,Y) or in the first return (Z_k, Z_{k+1}).

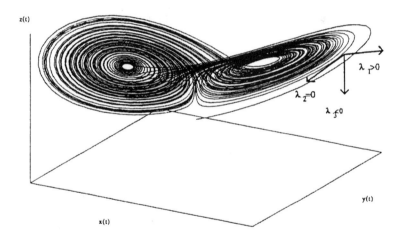

Figure 1.21. Lorenz attractor, with the parameters $P_r = 1.6$, $b = 4$, $r = 45.92$. From Hongre et al., 1994.

Hénon's Attractor

Hénon (1976) was trying to simplify the models of thermal convection when he replaced the system of 3 differential equations by an application with two dimensions. The points thus obtained in the plane can be considered as belonging to the Poincaré section of a 3-D flow. Hénon's map is:

$$\begin{cases} X_{k+1} = Y_k + 1 - \alpha X_k^2 \\ Y_{k+1} = \beta X_k \end{cases} \quad (1.56)$$

α and β are two constants respectively controlling the linearity and the dissipation of the system. The most commonly used values are $\alpha = 1.4$ and $\beta = 0.3$ (Figure 1.22).

The deterministic and chaotic character of Hénon's attractor is shown by starting from two points close to the attractor and distant from δ_0. The distance δ between the next points can be easily computed from the expressions of X_k and Y_k. δ increases with the iteration number k, following the law $\delta = \delta_0 \exp(\lambda_1 k)$. λ_1 is Lyapunov's largest exponent (see Section 1.5.6). Its value is $\lambda_1 = 0.5$. We find here the main properties of phase trajectories for a strange attractor, the SIC.

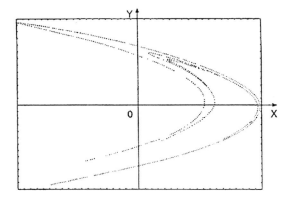

Figure 1.22. Hénon's Attractor. The attractor is shown here for $\alpha = 1.4$ and $\beta = 0.3$. The attractor's dimension is 1.26.

Another property observed earlier for the Lorenz attractor is the contraction of volumes or areas. The Lie derivative is similar to the Jacobian of the 2-D application:

$$J = \begin{vmatrix} \frac{\partial X_{k+1}}{\partial X_k} & \frac{\partial X_{k+1}}{\partial Y_k} \\ \frac{\partial Y_{k+1}}{\partial X_k} & \frac{\partial Y_{k+1}}{\partial Y_k} \end{vmatrix} = \begin{vmatrix} -2\alpha X_k & 1 \\ \beta & 0 \end{vmatrix} = -\beta \quad (1.57)$$

The areas defined by the phase trajectories are multiplied at each step by $|\beta|$. Here, we have $|\beta| < 1$ and therefore there is area contraction.

Ch. 1]   Limit Cycles and Attractors   47

Finally, the third of the properties mentioned above concerns the fractal structure of this attractor, whose dimension is not an integer. The attractor's structure repeats itself at different scales (Figure 1.23). The attractor's dimension is 1.26, and characterises an object between a line and a surface (Figure 1.23).

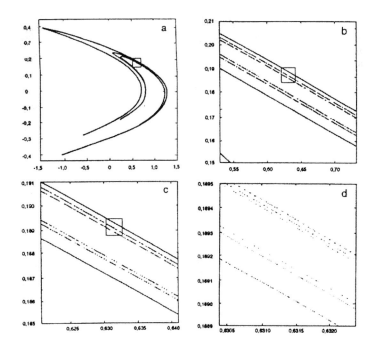

Figure 1.23. Hénon's Attractor. The structure repeats itself at different observation scales, showing its self-similarity. From Hénon, 1976.

Rössler's Attractor

The flow characterising Rössler's model has the shape:

$$\begin{cases} \dfrac{dX}{dT} = -(Y + Z) \\ \dfrac{dY}{dT} = X + aY \\ \dfrac{dZ}{dT} = b + XZ - cZ \end{cases} \quad (1.58)$$

When $a = b = 0.2$ and $c = 0.57$, the flow has a chaotic behaviour. If the values $b = 2$ and $c = 4$ are fixed, the nature of the attractor changes as a varies. When a is small, the attractor is a simple closed curve. When a increases, the attractor becomes a double loop, then a quadruple loop, and so on. When a = 0.375, the attractor is fractal and looks like a Moebius strip (Figure 1.24).

Rössler (1979) also defined the Rössler "hyperchaos" attractor:

$$\begin{cases} \dfrac{dX}{dT} = -(Y+Z) \\ \dfrac{dY}{dT} = X + aY + W \\ \dfrac{dZ}{dT} = b + XZ \\ \dfrac{dW}{dT} = cW - dZ \end{cases} \quad (1.59)$$

For a = 0.25, b = 3.0, c = 0.05, d = 0.5, the dimension of the attractor is 3.005.

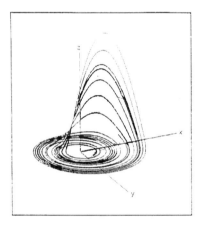

Figure 1.24. Rössler's Attractor. It is represented here in its phase space, and looks like a Moebius strip.

Feigenbaum's Attractor

Like Hénon's attractor, it is simply defined by an equation: $f_\lambda(x) = \lambda x (1-x)$. This attractor will be studied in better detail in the next chapter.

MacKey-Glass's Attractor

This attractor describes time series defined by the equation:

$$\frac{dX}{dt} = \frac{aX(t+s)}{1 + [X(t+s)]^c} - bX(t) \qquad (1.60)$$

This attractor has applications in biology to describe physiological control systems. For a = 0.2, b = 0.1, c = 10, and a given time delay s, the attractor's dimension is D = 6.7.

1.5.6 Lyapunov Exponents

The divergence of phase trajectories on the attractor was seen to be quite rapid. One parameter used to quantify the speed of this divergence is *Lyapunov's exponent*.

Let us first define the *Jacobian* of a transform. (Ox, Oy) and (Oz, Ot) are two coordinate systems, and there exists a linear transform which relates the point M, of coordinates x and y in the first system, to the point P of coordinates z and t in the second system (Figure 1.25), with:

$$\begin{cases} z = \alpha x + \beta y \\ t = \gamma x + \delta y \end{cases} \qquad \alpha; \beta; \gamma; \delta \text{ reals}$$

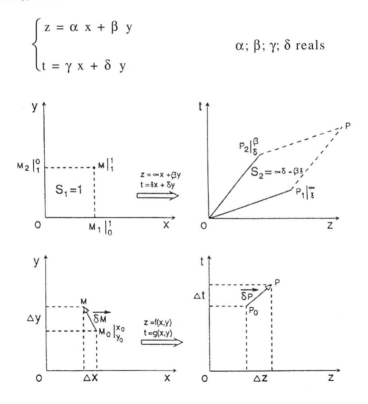

Figure 1.25. Linear and non-linear transforms. They respectively transform a point of the first system into a point of the second system, and a vector δ**M** close to M_0 into a vector δ**P**. The first transform shown is linear, the second is non-linear.

The two points $M_1(1,0)$ and $M_2(0,1)$ are respectively transformed into $P_1(\alpha,\gamma)$ and $P_2(\beta,\delta)$. The unit square OM_1MM_2 is transformed into a parallelogram OP_1PP_2, whose area is the absolute value of the determinant:

$$\begin{pmatrix} \alpha & \beta \\ \gamma & \delta \end{pmatrix} = \alpha\delta - \beta\gamma \qquad (1.61)$$

If $|\alpha\delta - \beta\gamma| > 1$, there is area expansion (dilation).
If $|\alpha\delta - \beta\gamma| < 1$, there is area contraction.

For a linear transform $z = f(x,y)$ and $t = g(x,y)$, close to $M_0(x_0,y_0)$ and for small increments Δx and Δy, it is possible to neglect the second-order terms and then:

$$df = \frac{\partial f}{\partial x} dx + \frac{\partial f}{\partial y} dy = f'_x\, dx + f'_y\, dy \qquad (1.62)$$

And therefore: $\Delta x \approx f'_x \Delta x + f'_y \Delta y$. The same goes for Δt.

If $\delta \mathbf{P}$ is the transform of $\delta \mathbf{M}$ in the neighbourhood of M_0, then:

$$\delta \mathbf{P} \approx \begin{pmatrix} f'_x(x_0,y_0) & f'_y(x_0,y_0) \\ g'_x(x_0,y_0) & g'_y(x_0,y_0) \end{pmatrix} \delta \mathbf{M} = J\, \delta \mathbf{M}$$

The matrix J is called the *Jacobian*, and one says that J is the Jacobian in (x_0,y_0) of the transform. If $J \neq 0$, the transformation can be locally inverted. If $|J| > 1$, there is area expansion around M_0. If $|J| < 1$, there is area contraction.

Let us now consider a map F in a p-dimensional space (Farmer et al., 1983): $x_{n+1} = F(x_n)$, where x is a p-dimensional vector.

$J(x) = \partial F/\partial x$ is the Jacobian of the application, and $J_n = [J(x_n), J(x_{n-1}), ..., J(x_1)]$. The eigenvalues of J_n are $j_1(n) \geq j_2(n) \geq ... \geq j_p(n)$. The Lyapunov numbers are:

$$\lambda_i = \lim_{n \to \infty} \left[j_i(n)^{1/n} \right], \; i = 1, 2, ..., p \qquad (1.64)$$

where only the positive n-th root is considered.

This definition was proposed by Oseledets (1968). By convention, $\lambda_1 \geq \lambda_2 \geq ... \lambda_n$.

In the case of a 2-D map, λ_1 and λ_2 are the axial lengths of an ellipse grown from an infinitesimally small circle. When the attractor is chaotic, statistically close points will diverge exponentially, and the Lyapunov numbers represent the half-axes of an ellipse (Figure 1.26). This translates quantitatively the notion of SIC.

The term (plural) of Lyapunov exponents is employed for the logarithms of the Lyapunov numbers.

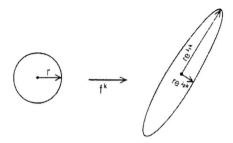

Figure 1.26. Two Lyapunov exponents. The exponents λ_1 and λ_2 are used to compute the lengths $re^{\lambda_1 k}$ and $re^{\lambda_2 k}$ of the ellipse formed after k iterations on the circle of radius r.

Figure 1.26 shows that n iterations of a 2-D map transform a small circle of radius δ into an ellipse with axes of respective lengths $\lambda_1^n \delta$ and $\lambda_2^n \delta$, with λ_1 and λ_2 being the Lyapunov numbers (Farmer et al., 1983). There is often confusion between the Lyapunov numbers and the Lyapunov coefficients. Following Falconer (1990), the general trend now is to note λ the exponents, remembering that after k iterations the circle of radius r forms an ellipse of axes $re^{\lambda_1 k}$ and $re^{\lambda_2 k}$ (Figure 1.26).

Practical computation

Let us consider a dynamic system observed with one of its time variables x(t), and let us suppose that the correlation function test of Grassberger and Procaccia showed this dynamic system to be chaotic and determinist, i.e. that it has an attractor in the phase space.

An m-dimensional phase picture is defined with a point on the attractor whose coordinates are $\{x(t), x(t+\tau), ..., x(t+(m-1)\tau)\}$, where τ is a time delay suitably chosen.

The initial point is $\{x(t_0), x(t_0+\tau), ..., x(t_0+(m-1)\tau)\}$, and the (Euclidean) distance to the closest point is named $L(t_0)$. At a later time t_1, the initial length has changed to $L'(t_1)$. If the time interval $[t_0, t_1]$ is too large, $L'(t_1)$ may be smaller than $L(t_0)$. This can happen for any interval in a portion of the attractor subjected to folding, and there is then under-estimation of λ_1. This is avoided by statistically analysing the $L'(t_k)$ on the whole series:

$$\lambda_1 = \frac{1}{t_M - t_0} \sum_{k=1}^{M} \log_2 \frac{L'(t_k)}{L'(t_{k-1})} \qquad (1.65)$$

where M is the total number of steps on the reference trajectory.

Figure 1.27 shows how to proceed. A reference trajectory[4] represents a point on the attractor at times $t_0, t_1, t_2, t_3, ..., t_k$. $L(t_0)$ is the distance at time t_0 between the point and its closest neighbour. At time t_1, it becomes $L'(t_1)$. $L(t_1)$ is the distance between the point at time t_1 and its closest neighbour, which becomes $L'(t_2)$ at time t_2, etc.

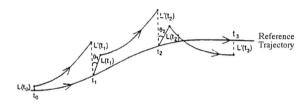

Figure 1.27. Lyapunov exponent. Representation in the phase space of the distances between trajectories, which are used to compute the largest exponent with the increase of the lengths L(t). From Wolf et al. (1985).

More information can be extracted from the series, and this method can be improved by computing the sum of the two largest Lyapunov exponents. The two closest neighbours in $t_0, t_1, t_2, ..., t_k$ define triangles of respective areas $A(t_0), A(t_1), A(t_2), ..., A(t_k)$. Thus $A(t_0)$ becomes $A'(t_1)$ at time t_1, and $A(t_1)$ becomes $A'(t_2)$, etc. (Figure 1.28). On the whole series:

$$\lambda_1 + \lambda_2 = \frac{1}{t_M - t_0} \sum_{k=1}^{M} \log_2 \frac{A'(t_k)}{A'(t_{k-1})} \tag{1.66}$$

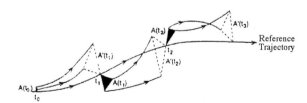

Figure 1.28. Method of the triangle areas. The triangles are constructed on the two closest neighbours, and the increase of their areas gives the two largest Lyapunov exponents (Wolf et al., 1985).

---

[4] also called *fiducial trajectory* by Wolf et al. (1985).

Kaplan and Yorke (1978) and Farmer et al. (1983) define the Lyapunov dimension, which in the case of a 2-D map equals the similarity dimension, but differs from the information dimension (which uses the probabilistic number of points in the unit square). We have the relation:

$$d_L = d_S = 1 + \frac{\log \lambda_1}{\log(1/\lambda_2)} \qquad (1.67)$$

It is therefore possible to compute the attractor's dimension, knowing λ_1 and λ_2.

In general:

$$d_L = k + \frac{\log(\lambda_1, \lambda_2, ..., \lambda_k)}{\log(1/\lambda_{k+1})} \qquad (1.68)$$

where k is the largest value for which $\lambda_1, \lambda_2, ..., \lambda_k \geq 1$.

If $\lambda_2 > 1$, $d_2 = 0$; if $\lambda_1, \lambda_2, ..., \lambda_k \geq 1$, $d_2 = p$.

This approach shows all its interest for computing an attractor's dimension using Lyapunov exponents, when the original series x(t) is short. For short series, indeed, Grassberger and Procaccia's method is not applicable. One *caveat* for the computation of λ_1 or $\lambda_1 + \lambda_2$ is that the series should not be too noisy. Wolf et al. (1985) investigated this problem and suggest several filters. One example is presented in Annex B.4, with the computation of Lyapunov exponents on a discrete series of geomagnetic declinations recorded at the Observatory of Chambon-la-Forêt, France.

1.5.7 Poincaré Sections and First-Return Maps

It is now easy to pass from dynamic systems to iterative processes and applications.

Let $\mathcal{D}$ be a subset of $\mathcal{R}^n$, and f: $\mathcal{D} \to \mathcal{D}$ a continuous map. f^k represents the k-th iteration: $f^0(x)=x$; $f^1(x)=f(x)$; $f^2(x)=f(f(x))$, etc. (Falconer, 1990). If x is a point in $\mathcal{D}$, then $f^k(x)$ is in $\mathcal{D}$, whatever the value of k. Usually, x, f(x), $f^2(x)$, ... are values of a given quantity at times 1, 2, ... Thus the value at time (k+1) is a function of the value at time k, using the function f. There are many examples of such systems: stable flow, discrete values of a physical parameter varying with time, etc.

An iterative system f^k is called a *discrete dynamic system*. One studies the behaviour of the series of iterations or orbits $\{f^k(x)\}_{k=1}^{\infty}$ for several initial points $x \in \mathcal{D}$ and for mainly high values of k. For example, if f(x)=cos.x, the series $f^k(x)$ tends toward 0.7391 when $k \to \infty$, whatever value of x was used at the beginning.

The distribution of iterations may seem random. Alternatively, $f^k(x)$ may tend toward a fixed point ω in D for which $f(\omega)=\omega$. More generally, $f^k(x)$ may tend toward an orbit of points with the period p (ω, f(ω), ..., f^{p-1}(ω)), where p is the smallest positive integer for which f^p(ω)=ω: $|f^k(x) - f^i(\omega)| \to 0$ when $k \to \infty$.

Sometimes, $f^k(x)$ may seem to vary randomly but remain close to a specific set which may be fractal. This brings us back to the notion of strange attractor. Actually, attractors could have been introduced that way, as sets toward which all neighbouring orbits are converging.

This is how the Hénon Attractor was defined earlier. The classical technique of studying an attractor of numerical series $f^k(x)$ consists in looking for the correlation function linking the points of the series in a phase space with a growing dimension p(x_k, x_{k+1}, ..., x_{k+p-1}) (Grassberger and Procaccia, 1983).

For a good look at this method we can come back to the flow's definition. Many dynamic systems are described by n first-order differential equations:

$$\frac{d\mathbf{X}}{dt} = \mathcal{F}(\mathbf{X},t) \qquad (1.69)$$

where $\mathbf{X}$ is a vector in the phases' space, $\mathcal{F}$ a function of $\mathbf{X}$ and of the time t. This system of equations defines a flow in $\mathcal{R}^n$ (Bergé et al., 1984).

To simplify, we take the case of n = 3, and a phase space with three dimensions x_1, x_2, x_3. To study this dynamic system in space, we look at the intersection of the phase trajectories with a plane S (e.g. the horizontal plane, x_3 = cst, of Figure 1.29). For a suitable value of this constant, the trajectory T cuts the plane S in several points P_0, P_1, P_2 (Figure 1.29). This set of points forms a Poincaré section, or first-return map.

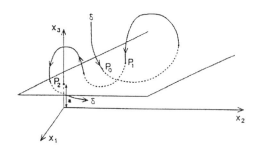

Figure 1.29. Poincaré section. It is defined here by the plane x_3 = a, which intersects a phase trajectory δ at points P_0, P_1, P_2 (descending trajectory).

The transform T that leads from one point to the next is a continuous map: $P_{k+1} = T(P_k) = T(T(P_{k-1})) = T^2(P_{k-1})$. If T is known, the point P_0 completely determines P_1, and P_2, P_k, etc. If P_1 determines P_0 when reversing time, T is said to be a reversible map from S to S.

If there is volume contraction in the phase space, this will be emphasised by area contraction in the plane S. If the flow has an attractor, its section will be visible in the plane S, and for an attractor of dimension d < 3, spatial representation will be achievable using several sections. The characteristics of the attractor can be found on the Poincaré section. The study of dynamic systems can therefore be simplified.

In the previous case, the Poincaré section allows one to pass from a flow in $\mathcal{R}^3$ to a map from the plane to itself. It also allows one to substitute algebraic equations for the differential equations governing the system. Iterating $x_i(k+1) = T(x_i(k))$ in the plane is easier than integrating a flow. Examples are the Poincaré section of a pendulum's limit cycle, which is a point, or the Poincaré section of a toroid, which is a circle. Figure 1.30 shows the Poincaré section of the Lorenz attractor in the plane $z = r-1$.

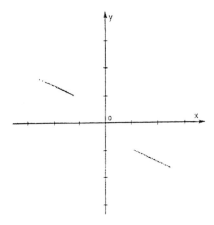

Figure 1.30. Poincaré section of the Lorenz attractor. The plane shown here is z = cst (cf. Figure 1.21).

These maps are therefore non-linear, mono-dimensional, of the type $x_{k+1}=f(x_k)$. They can describe chaotic behaviours. To better understand the interest of this approach, we will first illustrate it with a classic example, the quadratic or logistic map, Feigenbaum's Attractor (see Falconer, 1990; Bergé et al., 1984; Hentschel and Procaccia, 1983).

1.5.8 Logistic Map, Sub-Harmonic Cascade and Chaos

We will start from the continuous function (Bergé et al., 1984):

$$f(x) = 4 \mu x (1-x), \forall x \in [0,1], 0 \leq \mu \leq 1 \tag{1.70}$$

or: $f_\lambda(x) = \lambda x (1-x)$ (Falconer, 1990), to define a first-return map, which will for example be $x_{k+1} = 4 \mu x_k(1-x_k)$. This map relates any point x_k of the unit segment to another point x_{k+1} of the same segment. For a fixed value of $\mu = 0.5$, the graph of f is a parabola passing through $x = 0$, $x = 1$, and with a maximum at $x = 0.5$. We will plot the first bisecting line as well.

Although extremely simple, this function f and this first-return map create very complex situations when μ is varying. The map's nature will evidently depend on μ. Let us first look at the second intersection of the first bisecting line with the parabola f(x), and the slope of the local tangent to the curve. The intersection occurs at the point x verifying $x = 4\mu x (1-x) \Leftrightarrow x (= 4\mu x + 4\mu - 1) = 0$, i.e. $x = 0$ and $x = (4\mu - 1)/4\mu$. The local slope is $f'(x) = 4\mu (1 - 2x)$ and, at the second intersection, $f' = 2 - 4\mu$. And $f' = -1$ for a value $\mu_1 = 3/4$. The passing of f'(x) at -1 is characterised by an unstable behaviour of the map (see Figure 1.31).

1. $\mu < 0.75$

The graph (Figure 1.31a) shows that the iteration from any point x_0 different from 0 or 1 leads to the point x^*. This point is the intersection of the first bisecting line with the parabola (apart from 0).

This point, toward which any iteration starting from]0,1[converges, is stable; it is an attractor. This attractor is said to have a period of 1 (in the phases' space, this is indeed the period of the limit cycle, whose Poincaré section is this point).

Note that the absolute value of the slope at the intersection is smaller than 1.

2. μ slightly greater than 0.75

In this case, the absolute value of the slope is ≥ 1.

The attracting point becomes unstable (cf. Figure 1.31b, where $\mu = 0.8$). It is replaced by two points x_1^* and x_2^* which verify $x_2^* = f(x_1^*)$ and $x_1^* = f(x_2^*)$. The graph of $f^2(x)$ for $\mu = 0.8$ (Figure 1.31c) shows two stable points x_1^* and x_2^*, intersections of the bisecting line and $f^2(x)$. The map f(x) goes alternately from one to the other.

Starting from one point, two iterations are necessary to come back. These points form *an attractor of period 2*. On the limit cycle, two periods are necessary to come back to the same point of the Poincaré section.

To sum up, when m reaches the value of 0.75, an attractor of period 1 is replaced by an attractor of period 2; there is *bifurcation*.

3. μ increases

The graphs of f and f^2 are getting deformed (Figure 1.31d). The two points x_1^* and x_2^* become unstable when the absolute value of the tangent's slope on f^2 equals 1. A close-up view of f^2 around x_1^* and x_2^* shows a situation comparable to the previous case of x^* on f.

To each of these two points will be substituted two new points, i.e. a total of 4 for f^4. The critical value of μ for which this transition occurs is $\mu_2 = (1+ \sqrt{6})/4 = 0.86237...$

The attractor now encompasses 4 points, successively visited with a period of 4. Again, this period gets doubled with a *sub-harmonic bifurcation*.

The process can be repeated *ad infinitum* by augmenting the values of μ, creating a *cascade of bifurcations* (Figure 1.32).

This reasoning shows an analogy of structures at all scales, like the Cantor set. When μ increases, there is a succession of attractors of period 2^l, l integer varying from 0 (for $\mu_1 \leq 0.75$) to ∞. It is numerically possible to show that $\mu_\infty = 0.892486418...$

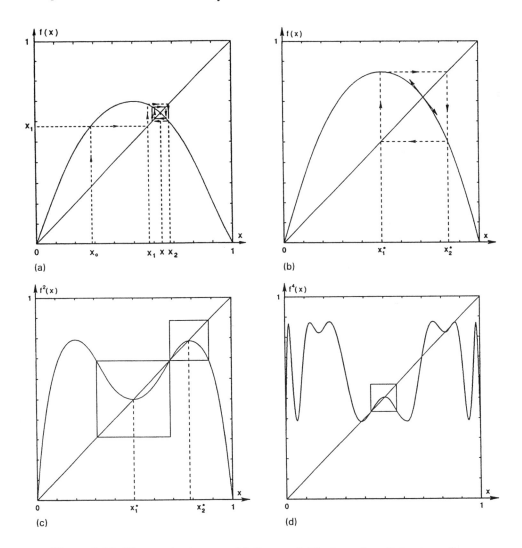

Figure 1.31. The quadratic map; (a) for $\mu < 0.75$, the point of abscissa x^* and corresponding to the intersection of the first bisecting line with the graph of $f(x)$ is an attractor toward which all iterations converge; (b) for $\mu = 0.8$, the attracting point is replaced by x_1^* and x_2^* which verify $x_2^* = f(x_1^*)$ and $x_1^* = f(x_2^*)$; (c) for $\mu = 0.8$, and on the graph of $g(x) = f(f(x)) = f^2(x)$, the two points x_1^* and x_2^* are attractors. Points alternate from one attractor to the other. The structure of each of the outlined squares is identical to the structure of (b); (d) for $\mu = 0.875$, this is the graph of $h(x) = g(g(x)) = f^4(x)$. There are four stable attracting points. The outlined square is similar to (b).

The convergence of μ_∞ follows a scaling law:

$$\lim_{i \to \infty} \frac{\mu_i - \mu_{i-1}}{\mu_{i+1} - \mu_i} = \delta \qquad (1.71)$$

The scale reduction factor δ is a universal number independent from the actual definition of f, and $\delta = 4.6692016091029909...$

On the segment [0,1], the scale reduction factor separating the points of the Poincaré section is $\alpha = 2.502907...$ (cf. Feigenbaum's Attractor).

For values of μ greater than $\mu_\infty = 0.892...$, one enters a region where aperiodic and periodic attractors alternate (Figure 2.35). For aperiodic attractors, this is chaos, where the iterations of f yield successive values of x that never repeat themselves and which depend on the initial conditions x_0. This is the SIC described earlier.

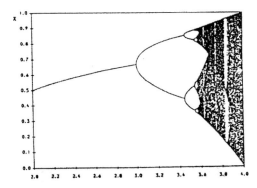

Figure 1.32. A cascade of bifurcations. When μ increases, a series of sub-harmonic bifurcations can be observed, the number of fixed points is each time multiplied by two.

Note: the spelling of Lyapunov's name is quite varied, and one finds the four following cases:
- Liapunov, in: Eckman, J.P., S. Oliffson-Kamphorst, D. Ruelle, S. Ciliberto; "Liapunov exponents from time series", Phys. Rev. A, vol. 34, no. 6, p. 4971-4979, 1986
- Lyapunov: Farmer, J.D., E. Ott, J.A. Yorke; "The dimension of chaotic attractors", Physica D, vol. 7, p. 153-180, 1983; Wolf, A. J.B. Swift, H.L. Swinney, J.A. Vastano; "Determining Lyapunov exponents from a time series", Physica D, vol. 16, p. 285-317, 1985; Bergé, P., Y. Pomeau, C. Vidal; "L'ordre dans le chaos: vers une approche déterministe de la turbulence", Hermann: Paris, 353 pp., 1984

- Lyapounov: Prigogine, I., T.Y. Petrovsky, H.H. Hasegawa, S. Tasaki; "Integrability and chaos in classical and quantum mechanics", Chaos, Solitons and Fractals, vol. 1, no. 1, p. 3-24, 1991
- Liapounov: Falconer, K.; "Fractal geometry: mathematical foundations and applications", Wiley: Chichester, 288 pp., 1990

The original spelling is found in Cyrillic in the thesis of the Russian mathematician Valery Oseledets (1967). His first English publication (Oseledets, V.; "Multiplicative ergodic theorem - Lyapunov characteristic numbers for dynamic systems", Transactions of the Moscow Mathematical Society, vol. 19, p. 197-231, 1968) uses the second spelling, which we reproduce here.

1.6 MULTIFRACTALS AND WAVELET TRANSFORMS

Let us suppose that a regional mass distribution μ varies greatly. If this mass concentration has a certain mass density, e.g. $\mu(B_r(x)) = r^\alpha$ for r small, and different subsets will have different values of α, the mass distribution or measure μ is called a *multifractal measure*. Before examining more rigorously this definition, we will define the main dimensions that have been proposed to characterise fractal sets.

1.6.1 Notion of Generalised Dimension

The studies of Hentschel and Procaccia (1983), following those of Takens (1980) and Grassberger (1981, 1983), have clarified this notion, and proved there was an infinite number of generalised dimensions D_q, q being a strictly positive integer, which can characterise fractal sets or strange attractors.

Three different dimensions were first defined; the similarity dimension D, the information dimension σ, and the correlation dimension ν. To explicitly define them, we will consider a long time series $[X_i]_{i=1}^N$ where N is very large but finite, on an attractor. The phase space will be covered with a grid of cubes of (Euclidean) dimension d, of sizes b^d. M(b) is the number of cubes containing points of the series $[X_i]_{i=1}^N$; N_k is the number of points in the cube of order k, and $P_k \equiv N_k / N$.

<u>Similarity Dimension</u>

It is defined by:

$$D = \lim_{b \to 0} \lim_{N \to \infty} \frac{\log.M(b)}{\log.b} \tag{1.72}$$

This is the dimension defined by Mandelbrot. This is also the definition we obtained with the method of Cantor's Dust (see Section 1.3).

A more detailed description of the fractal object studied (here, the attractor of a dynamic system in its phase space) can be obtained by using a dimension holding more information. This is the information dimension.

Information Dimension

It is expressed as:

$$\sigma = - \lim_{b \to 0} \lim_{N \to \infty} \frac{\log.S(b)}{\log.b} \tag{1.73}$$

where the system's entropy is:

$$S(b) = - \sum_{k=1}^{M(b)} P_k \log.P_k \tag{1.74}$$

Correlation Dimension

This dimension was defined earlier, and is expressed as:

$$\nu = \lim_{b \to 0} \lim_{N \to \infty} \frac{C(b)}{\log.b} \tag{1.75}$$

where:

$$C(b) = \frac{1}{N^2} \sum_{i \neq j} H(b - |X_i - X_j|) \tag{1.76}$$

and H is the Heaviside function (see Section 1.5.4). The practical computation of this dimension was already presented: the dimension of the phase space is continuously augmented until saturation of C(b), and this integer Euclidean dimension is called the *embedding dimension*. This dimension only characterises the phase space, and should not be confused with the dimension of the attractor.

Hentschel and Procaccia (1983) demonstrated there was a hierarchy of generalised dimensions D_q.

Generalised Dimensions - Rényi Dimensions

The generalised dimensions are defined for any $q \geq 1$. When q is an integer, D_q has a physical meaning.

$$D_q = \frac{1}{q-1} \lim_{b \to 0} \left(\frac{\log. \sum_i p_i^q}{\log.b} \right) \quad (1.77)$$

where p_i is the probability of finding points in the box of size b.

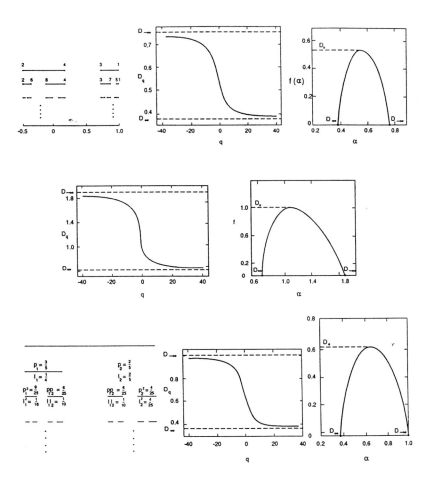

Figure 1.33. Notion of generalised dimension. For q varying between $-\infty$ and $+\infty$, the values corresponding to the singularity spectra are plotted for three different sets: period-doubling, quasi-periodic circle of trajectories, Cantor's multifractal. From Halsey et al. (1986).

Hentschel and Procaccia (1983) demonstrated that:

$$\begin{cases} D = \lim_{q \to 0} D_q \\ \sigma = \lim_{q \to 1} D_q \\ \nu = \lim_{q \to 2} D_q \end{cases} \quad (1.78)$$

For $q = 3, 4, ..., r$, the dimensions D_q are associated to the correlation integrals of the triplets, quadruplets, ... of points on the attractor. Furthermore:

$$D_q > D_{q'}, \forall\, q' > q \quad (1.79)$$

In the case of Feigenbaum's Attractor, $D = 0.537$; $\sigma = 0.518$; $\nu = 0.501$; $D_\infty = 0.394$.

This technique can be extended to negative values of q (Figure 1.33). On the graph of $D_q = f(q)$, there is symmetry between the asymptotic values of $D_{-\infty}$ and D_∞.

1.6.2 Multifractals - Theory and Practice

The notion of multifractals was first introduced by Frisch and Parisi (1985) to better describe some problems encountered in turbulence studies. Mandelbrot (1989) gives a simple definition of multifractals when he writes: "*the generalisation of fractals to multifractals corresponds to the transition from geometrical objects characterised first by a number, to geometrical objects characterised first by a function which can be a probability distribution*".

Geilikman et al. (1990) present this simply. Let N be the number of boxes of sizes Δ necessary to cover a particular set. For this fractal set, $N \sim \Delta^{-D_0}$ when $\Delta \to 0$. The exponent D_0 is considered as the set's dimension, and may not be an integer (this is a fractal set). Let us now look at the probability $p_i(\Delta)$ for a point x_i of the set to be included in a box of size Δ. For a simple fractal set, $p_i(\Delta) = \Delta^{D_0}$ independently from the point x_i considered. For more complex sets, $p_i(\Delta) = \Delta^{\alpha(x_i)}$ with the exponent $\alpha(x_i)$ different from one point to another. $\alpha(x_i)$ is called the *singularity index* or punctual dimension of the set. It is defined by:

$$\alpha_i = \alpha(x_i) = \frac{\log.p_i(\Delta)}{\log.\Delta}, \Delta \to 0 \quad (1.81)$$

This forms a spectrum of different dimensions, created by a non-uniform distribution of probabilities for the possible points. For multifractal sets, it is not unreasonable to assume that the number of boxes with $\alpha(x_i)$ in the interval $[\alpha, \alpha+\delta\alpha]$ is given by:

$$N_\Delta\, \delta\alpha \approx \rho(\alpha,\Delta)\, \Delta^{-f(\alpha)\delta\alpha}, \Delta \to 0 \quad (1.81)$$

where $\rho(\alpha,\Delta)$ is a function of δ. The function $f(x)$ describes the spectrum of singularity indices. Using the dimension D_q defined earlier (section 1.6.1), we can get:

$$D_q = \lim_{\Delta \to 0} \left(\frac{\log. \sum_i \log.p_i(\Delta)^q}{(q-1) \log.\Delta} \right)$$

where the integer q is the moment of a probability distribution.

To estimate $\sum_i$, Geilikman et al. (1990) show that:

$$f(\alpha_q) = q\,\alpha_q - (q-1)\,D_q \qquad (1.82)$$

$$\alpha_q = \frac{d}{dq}[(q-1)\,D_q] \qquad (1.83)$$

After elimination of q, these equations give f(α). The functions f(α) and (q - 1) D_q are the **Legendre Transforms** of one another.

Falconer (1990) gives an interesting theoretical example of a multifractal set built upon the classic triadic Cantor set.

It starts from the set $\mathcal{F} = \bigcap_{k=0}^{\infty} \varepsilon_k$, ε_k containing 2^k intervals of length 3^{-k}.

Choosing $0 < p < 1$, it is possible to define a measure μ. The mass unit is divided to get a mass of p for the left third of ε_1, and a mass of (1-p) for the right third of ε_1. The mass of each interval of ε_1 is then divided by p / (1-p) between the two sub-intervals of ε_2. The process is repeated at each iteration, i.e. the mass of each interval of ε_k is divided by p/(1-p) between the two resulting sub-intervals in ε_{k+1} (Figure 1.34).

The sum of the masses of all the filled intervals is always 1. A mass distribution μ has therefore been defined on $\mathcal{F}$.

For any $0 < \delta < 1$, if ($\mathcal{B}_i$) are the cubes of coordinate δ which cover the base of μ, we can count the number of cubes δ for a "reasonably large" measure (i.e. mathematically different from 0, though $k \to \infty$).

For $-\infty < \alpha < +\infty$, $N_\delta(\alpha)$ follows a power law, and we can compute $\log.N_\delta(\alpha)/\log.\delta$ when $\delta \to 0$. We can also compute (and this is new) the sum of the masses in the cubes δ:

$$S_\delta(q) = \sum_i \mu(\mathcal{B}_i)^q \qquad (1.84)$$

For $-\infty < q < +\infty$, i is the number of cubes necessary to cover the base of μ. Note that:
$S_\delta(0) = N_\delta$ (base of μ)

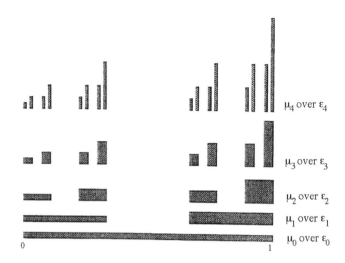

Figure 1.34. A multifractal set built upon a triadic Cantor set. The sum of all masses (or probabilities) is always equal to 1 (Falconer, 1990).

In our example, ε_k is made up of 2^k intervals of length 3^{-k}, and for any r a number $\binom{k}{r}$ of these intervals has a mass $p^r(1-p)^{k-r}$. $\binom{k}{r}$ is the binomial distribution coefficient:

$$\binom{k}{r} = C_r^k = \frac{k!}{r!\,(k-r)!} \tag{1.85}$$

Figure 1.34 shows the multifractal construction for $p = 1/3$ (i.e. a ratio $p/(1-p) = 1/2$).

The theorem of the binomial distribution gives:

$$S_{3^{-k}}(q) = \sum_{r=0}^{k} \binom{k}{r} p^{qr}(1-p)^{q(k-r)} = (p^q + (1-p)^q)^k \tag{1.86}$$

This can be empirically verified on Figure 1.34, and yields:

$$S_\delta(q) = \delta^{\log.(p^q + (1-p)^q)/\log.3} \tag{1.87}$$

And:

$$\lim_{\delta \to 0} \frac{\log.S_\delta(q)}{-\log.\delta} = \log.\frac{p^q + (1-p)^q}{\log.3} \tag{1.88}$$

Evaluating $N_\delta(\alpha)$ is even more difficult:

$$\lim_{\varepsilon \to 0} \lim_{\delta \to 0} \frac{\log.(N_\delta(\alpha+\varepsilon)-N_\delta(\alpha-\varepsilon))}{-\log.\delta} \equiv f(x) \tag{1.89}$$

A rather laborious computation (Falconer, 1990) leads to:

$$f(\alpha(q)) = \frac{\left(\log(p^q + (1-p)^q) - \dfrac{q(p^q\log.p+(1-p)^q\log.(1-p))}{(p^q + (1-p)^q)}\right)}{\log.3} \tag{1.90}$$

$f(\alpha)$ can therefore be computed as a function of the singularity index α (Figure 1.35). It corresponds to the spectrum of singularity indices for the multifractal set considered. And, of course, for q=0, $f(\alpha) = \dfrac{\log.2}{\log.3}$.

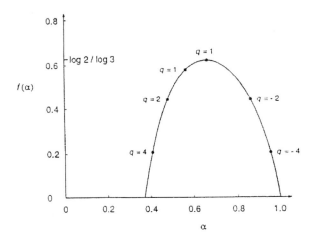

Figure 1.35. Example of a singularity spectrum. This is the multifractal spectrum of the Cantor set shown in Figure 1.34, for q varying between $-\infty$ and $+\infty$

Practice

As we mentioned at the beginning of this section, a fractal set cannot usually be described by a single dimension. It is therefore worth looking for a spectrum of singularity indices, i.e. describing quantitatively a multifractal set. The sets dealt with generally are time series of a variable characteristic of a dynamic system. As in section 1.3.8, where the correlation dimension of an attractor was introduced for a series of time values, we have a series of $[x(t_i)]_{i=0}^{N}$, t_i defined with a step τ, for which we can build a phase space. The series can also be defined by the distance between two successive events (volcanic eruptions, earthquakes, geomagnetic space, etc.). The phases' space can then be built on the order of these events.

Let us first look at a first-return map X_{n+1}, X_n, and at the scale factors α and $f(\alpha)$ presented earlier, also defined by Halsey et al. (1986).

$p_i^q \sim l_i^{\alpha q}$, where $p_i = N_i/N$ is the probability for a point in the phase space to be inside the box (or target) of size l_i, and where the integer q is similar to the moment of a probability distribution:

$$n(\alpha,l) = d\alpha' \, \rho(\alpha') \, l^{-f(\alpha')} \qquad (1.91)$$

where $n(\alpha,l)$ is the number of times α is found in the interval $d\alpha'$ around α'. As α' may be found at several places on the attractor, it makes up a subset with a proportionality factor $\rho(\alpha')$, a fractal dimension $f(\alpha')$ and a density $l^{-f(\alpha')}$.

The distribution function is $\Gamma = \sum_i (P_i)^q/l_i^\tau$, where τ is a function of q, α and $f(\alpha)$.

Halsey et al. (1986) and Halsey and Jensen (1986) showed that the distribution function leads to the following relations:

$$\begin{cases} \tau = \alpha q - f(\alpha) \\ \alpha = \dfrac{d\tau}{dq} \\ \dfrac{df}{d\alpha} = q \text{ and } \dfrac{d^2 f}{d^2 \alpha} < 0 \end{cases} \qquad (1.92)$$

These equations are similar to the Legendre transforms used in thermodynamics, where $f(\alpha)$ is analogous to entropy, α to the energy, and τ to the free energy.

Following the method of Jensen et al. (1985), the recurrence probability m_i can be estimated practically with $m_i = p_i^{-1}$, where m_i is the number of steps necessary to come back inside the circle of radius l centred on the starting point (Figure 1.36).

The values of m_i are determined, for one value of l, on all points of the series. They are then elevated to the power (1-q) and averaged. This is repeated for several values of l and q. For a constant q, τ is the slope of the log-log graph of the averaged (m^{1-q}) vs. l. This allows one to calculate $\alpha = d\tau/dq$ and $f = q\alpha - \tau$.

This produces the multifractal spectrum $f(\alpha)$ of the singularities (see Figure 1.35). The curves $f(\alpha)$ have maxima between α_{min} and α_{max}. $f(\alpha=\alpha_{min}) = f(\alpha=\alpha_{max}) = 0$. $f(\alpha)_{max}$ is the dimension of the whole attractor (Halsey et al., 1986).

This very laborious technique can be easily programmed on computers, and describes perfectly the multifractal appearance of an attractor. It was applied successfully to the volcanic tremors of Hawaii (Shaw and Chouet, 1989) (see Chapter 2), under the name of F.S.A. (Fractal Singularity Analysis).

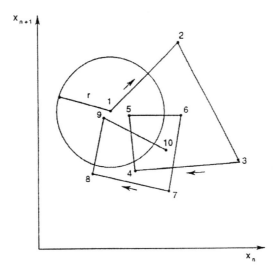

Figure 1.36. Practical computation of a singularity spectrum. Described in the text, Jensen's method consists in systematically counting for each point of the first-return map the probability of time recurrence to come back inside the circle centred on this particular point.

1.6.3 Wavelet Transform

The generalised fractal dimensions D_q and the singularity spectrum $f(\alpha)$ only provide a statistical description of the respective contributions of each singularity. The additional information about their spatial location is obtained by extending the wavelet analysis of fractal functions to measures distributed on Cantor sets.

Wavelet analysis is a mathematical technique recently introduced to analyse seismic data and acoustic signals. It unfolds 1-D signals into 2-D, using scale and location as independent variables (Argoul et al., 1989).

The method consists in expanding an arbitrary real function $S(x)$ onto wavelets constructed with a simple function g, with dilations and translations. The wavelet transform of $S(x)$ by the wavelet g is defined by:

$$\mathcal{T}_g(a,x) = \omega(a) \int \left(\frac{g(x-y)}{a}\right) S(y)\, dy \qquad a > 0,\, x \in \mathcal{R} \qquad (1.93)$$

where the weight function ω(a) represents the visual enlargement. The wavelet g is a regular function, localised around x=0.

For a large class of functions S(y), the wavelet transform can be inverted, provided g satisfies a certain number of conditions, including:

$$\int g(y)dy = 0 \tag{1.94}$$

The wavelet transform can be considered as a *mathematical microscope*. The location and the enlargement factor correspond to x and a^{-1}, and the optical performance is determined by the choice of the analysing wavelet.

Wavelet transforms has become a powerful tool for locating singularities: a singularity of S(x) in x_0 produces a cone-shaped structure in $\mathcal{T}_g$, pointing from (a=0, x=x_0). The nature of singularities can be deduced from the power-law behaviour of $\mathcal{T}_g(a, x=x_0)$ by increasing the enlargement factor a^{-1}.

Arnéodo et al. (1988) demonstrated that the wavelet transform helps to visualise the self-similar properties of fractal objects. Specifically, it shows the complexity of the fractal object observed, revealing the hierarchy of the relative positions of singularities. Figure 1.37a shows the successive branches and forks observed on $\mathcal{T}_g(a, x)$ when the enlargement factor a^{-1} increases on the simple triadic Cantor set (single scale). Figure 1.37b shows the same analysis performed on a multifractal Cantor set (double scale); the branches and forks are not symmetrical anymore when a^{-1} increases, they do not occur in pairs anymore, but successively, as the signature of two different scale factors.

Among the functions most commonly used for g, we can cite:

<u>The Morlet Wavelet</u>

Historically, this is the first wavelet to have been defined (Grossman and Morlet, 1984). It is made of superposed Gaussians centred around the frequency ω.

For any $\omega \in \mathcal{R}$, the functions:

$$g_\Omega(\omega) = e^{-(\omega-\Omega)^2/2} - e^{-\Omega^2/4} e^{-(\omega-\Omega)^2/4} \tag{1.95}$$

satisfy the condition $g_\Omega(0) = 0$.

The functions g_Ω, computed by inverse Fourier transform, are also wavelets:

$$g_\Omega(t) = e^{i\Omega t} e^{-t^2/2} - \sqrt{2} e^{-\Omega^2/4} e^{i\Omega t} e^{-t^2} \tag{1.96}$$

This family of wavelets with a single parameter Ω form the *Morlet wavelets*. For a value of Ω large enough, the rightmost term of equation (1.96) becomes negligible.

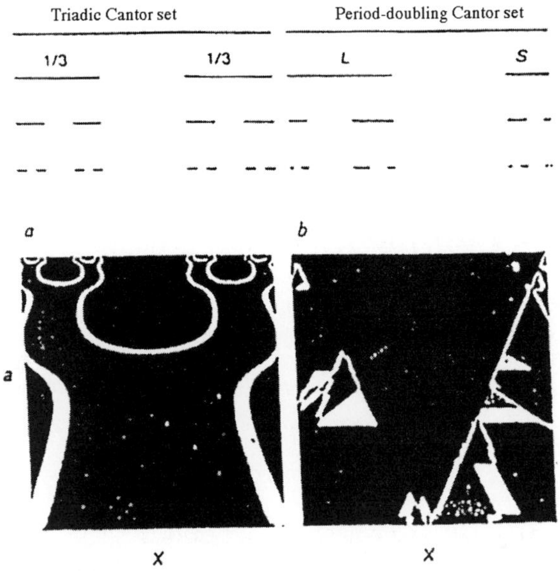

Ch. 1]   Multifractals and Wavelets   69

Figure 1.37. Wavelet analysis of Cantor sets with a Mexican Hat. The x-axis corresponds to the space where the sets are constructed, the y-axis corresponds to the enlargement factor a (Argoul et al., 1989).

The Mexican Hat

This wavelet is the second derivative of the Gaussian and is often used:

$$g(t) = (1 - t^2) e^{-t^2/2} \tag{1.97}$$

Its Fourier transform becomes 0 for $\omega = 0$:

$$g(\omega) = \omega^2 e^{-\omega^2/2} \tag{1.98}$$

The wavelet is well localised in both spaces, and therefore proves useful for locating singularities in multifractal sets.

Wavelets constant by parts

First used by computer scientists, they can easily represent the Mexican hat in algorithms. For example:

$$g(t) = \begin{cases} 1 & \text{for } |t| < 1 \\ -1/2 & \text{for } 1 < |t| < 3 \\ 0 & \text{for } |t| > 3 \end{cases} \tag{1.99}$$

Their application to multifractal sets:

As an example, we will consider a multifractal set in which a measure $\mu(I(x,\varepsilon))$ for an interval I is centred on point x and increases with its length ε:

$$\mu(I(x,\varepsilon)) = \int_{I(x,\varepsilon)} d\mu(g) \sim e^D \qquad (1.100)$$

For the wavelet analysis of the measure μ, the definition of the wavelet transform is extended to:

$$\mathcal{T}_g(a,x) = \frac{1}{a^{-n}} \int g^*\left(\frac{x-b}{a}\right) d\mu(x) \qquad (1.101)$$

The normalisation factor a^{-n} allows one to optimise the visualisation of μ's fractal structure. Indeed, by taking $n > \alpha_{max}$, (α being the exponent of the singularity spectrum), each singularity of μ will manifest iself by a power-law divergence of $\mathcal{T}_g$, when $\alpha \to 0^+$.

If μ shows a scaling behaviour in the neighbourhood of x_0, i.e. if :

$$\mu(I(x_0,\lambda\varepsilon)) \sim \lambda^{\alpha(x_0)} \mu(I(x_0,\varepsilon)) \qquad (1.102)$$

for integer exponents $\alpha(x_1)$ and an analysis wavelet rapidly decreasing at ∞, the local scaling behaviour of the measure μ is reflected in its wavelet transform:

$$\mathcal{T}_g(\lambda a, x_0+\lambda b) = (\lambda a)^{-n} \int g^*\left(\frac{x-x_0-\lambda b}{\lambda b}\right) d\mu(x) \qquad (1.103)$$

$$= (\lambda a)^{-n} \int g^*\left(\frac{x-\lambda b}{\lambda a}\right) d\mu_{x_0}(x)$$

$$= (\lambda a)^{-n} \int g^*\left(\frac{y-b}{a}\right) d\mu_{x_0}(\lambda y)$$

Hence:

$$\mathcal{T}_g(\lambda a, x_0+\lambda b) \sim \lambda^{\alpha(x_0)-n} \mathcal{T}_g(a, x_0+b) \qquad (1.104)$$

When $\lambda \to 0$, $\mathcal{T}_g$ behaves as a power law, with an exponent $\alpha(x_1) = \alpha(x_0)-n$.

Figure 1.38 shows the results obtained by Arnéodo et al. (1988), using a Gaussian wavelet transform on a uniform Cantor set, with $p_1 = p_2 = 1/2$, and on a multifractal measure generated with $p_1 = 3/4$, $p_2 = 1/4$.

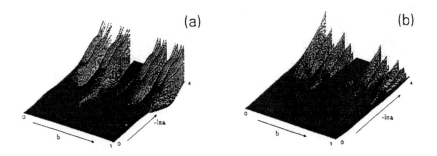

Figure 1.38. Two wavelet transforms of a triadic Cantor set. The transform is $(\text{sgn}(\mathcal{T}) | \mathcal{T}(a,b)|^{1/2})$ with: (a) a uniform measure; (b) two distinct measures $p_1 = 3/4$, $p_2 = 1/4$. From Arnéodo et al., 1988.

1.7. NOTION OF CHAOS

Even in the greatest order,
there is a small disorder
G.W. Leibniz

The path to chaos was first mentioned when studying the first-return map (see Section 1.5), when looking in detail at the quadratic map where the parameter μ is varying. We had observed a series of bifurcations and the apparition of chaos when μ fell between 0.892 and 1. The present section will study the problem in general.

1.7.1 The Path to Chaos - Hopf Bifurcation

For a simple periodic system, the phase trajectory was seen to tend toward a limit cycle, whose Poincaré section was reduced to a single point P_0. It is a fixed point of the mapping:

$$\mathcal{T}: P_0 = \mathcal{T}(P_0) = \mathcal{T}^2(P_0) = \ldots \tag{1.105}$$

There are three possibilities for losing the stability of this periodic solution. A linear analysis with only the first-order terms is enough: the map $\mathcal{T}$ is defined, at a first order, by the Floquet matrix $\mathcal{M}$. In the neighbourhood of O:

$$\mathcal{M} = [\partial T / \partial x_i]_{x_i, 0} \tag{1.106}$$

Let us first examine the properties of this matrix.

<u>Floquet Matrix</u>

P_0 is the stable point of the Poincaré section corresponding to the stable limit cycle. P_0, P_1, P_2, ..., P_n are the points where the trajectory intersects the Poincaré section, close to the limit cycle. We will find out how P_2, P_3, ..., P_n are deduced from P_0 (Bergé et al., 1984). A system of axes P_0x, P_0y can be defined in the section's plan. P_1 has $\{x_1, y_1\}$ for coordinates, P_2 has $\{x_2, y_2\}$, etc. (Figure 1.39).

The linear transformation defining the coordinates of P_2, image of P_1, is:

$$P_2 = \begin{pmatrix} x_2 = ax_1 + by_1 \\ y_2 = cx_1 + dy_1 \end{pmatrix} \tag{1.107}$$

and:

$$P_2 = \begin{pmatrix} x_n = ax_{n-1} + by_{n-1} \\ y_n = cx_{n-1} + dy_{n-1} \end{pmatrix} \tag{1.108}$$

The transformation matrix is therefore:

$$\mathcal{M} = \begin{pmatrix} a & b \\ c & d \end{pmatrix} \tag{1.109}$$

We have:

$$\begin{pmatrix} x_n \\ y_n \end{pmatrix} = \mathcal{M}^{n-1} \begin{pmatrix} x_1 \\ y_1 \end{pmatrix} \tag{1.110}$$

That is:

$$\mathbf{P_0 P_n} = \mathcal{M}^{n-1} \mathbf{P_0 P_1} \tag{1.111}$$

Computing $\mathcal{M}^{n-1}$ is laborious. To simplify the transformation, we are looking for a system of coordinates x'y' and two numbers λ_1, λ_2 such that P_2 can be deduced from P_1 by:

$$P_2 = \begin{cases} x_2' = \lambda_1 x_1' \\ y_2' = \lambda_2 y_2' \end{cases} \tag{1.112}$$

and

$$P_n = \begin{cases} x_n' = \lambda_1^{n-1} x_1' \\ y_n' = \lambda_2^{n-1} y_2' \end{cases} \quad (1.113)$$

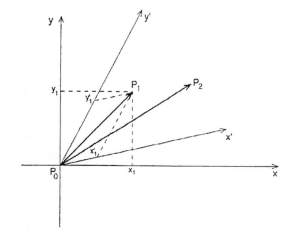

Figure 1.39. Floquet matrix. Here are represented the eigendirections P_0x' and P_0y' of the Floquet matrix.

The new transformation matrix $\mathcal{M}$ is diagonal:

$$\mathcal{M} = \begin{pmatrix} \lambda_1 & 0 \\ 0 & \lambda_2 \end{pmatrix} \quad (1.114)$$

and:

$$\mathcal{M}^{n-1} = \begin{pmatrix} \lambda_1^{n-1} & 0 \\ 0 & \lambda_2^{n-1} \end{pmatrix} \quad (1.115)$$

We can recognise the eigenvalues λ_1 and λ_2 of $\mathcal{M}$, which are the roots of the equation:

$$\begin{pmatrix} a-\lambda & b \\ c & d-\lambda \end{pmatrix} = 0 \implies \lambda^2 - (a+d)\lambda + (ad-bc) = 0 \quad (1.116)$$

These roots may be imaginary and conjugate: $\lambda_1, \lambda_2 = \alpha \pm i\beta$. The eigenvectors $\mathbf{V}$ of $\mathcal{M}$ are such that $\mathcal{M}.\mathbf{V} = \lambda \mathbf{V}$. They are co-linear to the axes P_0x and P_0y; these are the eigendirections of $\mathcal{M}$. They are defined, in x0y, by:

$$\begin{cases} ax + by = \lambda x \\ cx + dy = \lambda y \end{cases} \quad (1.117)$$

As:

$$P_n = \begin{cases} x_n' = \lambda_1^{n-1} x_1' \\ y_n' = \lambda_2^{n-1} y_2' \end{cases}$$

it is the largest module $|\lambda|^+$ of λ (respective to 1) which influences the evolution of P_n.

$$\begin{cases} \text{When } |\lambda|^+ < 1 \quad \lambda^n \to 0 \text{ and } P_n \to P_0 \\ \text{When } |\lambda|^+ > 1 \quad \lambda^n \to \infty \text{ and } P_n \to \infty \end{cases} \quad (1.118)$$

This expression of the Floquet matrix means that, after one cycle, the transform P_1 of P_0 is $P_0+\delta$ and $\mathcal{T}(P_0+\delta) - P_0 \approx \mathcal{M}\delta$, with $|\delta| \to 0$.

How stable the trajectory is, will depend on the eigenvalues of the Floquet matrix, as after m cycles: $\mathcal{T}^m(P_0+\delta) - P_0 \approx \mathcal{M}^m \delta$,. The gap will tend exponentially towards 0 if the eigenvalues of $\mathcal{M}$ are strictly less than 1. If at least one of the eigenvalues of $\mathcal{M}$ is greater than 1, the gap will augment with time and the cycle is unstable.

The important result about the properties of the Floquet matrix can therefore be stated as: "the loss of stability of the limit cycle corresponds to the crossing of the unit circle by at least one of the eigenvalues of the Floquet matrix".

<u>Hopf Bifurcation</u>

Let us now examine how a periodic attractor becomes a strange attractor. During the loss of stability of a periodic system, three paths are possible.

On Figure 1.40, the loss of stability is shown by:

$$|x_0x_1| < |x_0x_2| < \ldots \quad (1.119)$$

If $\delta x_1 = x_0x_1$ and $\delta x_2 = x_0x_2$, we have

$$\delta x_2 = \mathcal{M} \delta x_1 \quad (1.120)$$

Ch. 1]   Notion of Chaos   75

λ is the eigenvalue of the Floquet matrix.

(1) if $\lambda = 1+\varepsilon$, δx_i increases at each iteration

(2) if $\lambda = -(1+\varepsilon)$, the module of δx_i increases at each iteration but its direction points alternatively in two opposite directions.

(3) if λ is a complex conjugate, $\lambda = \alpha \pm i\beta$ with $|\lambda| > 1$, the successive δx_i rotate with an angle γ at each iteration, and the module of δx_i increases (Figure 1.40).

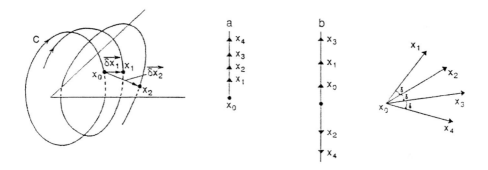

Figure 1.40. Poincaré section of a phase trajectory. C is the limit cycle which crosses the section in x_0. In the plane of the Poincaré section, the three possibilities are represented for: $\lambda > 1$; $\lambda < -1$; $\lambda = \alpha \pm i\beta$ and $|\lambda| > 1$.

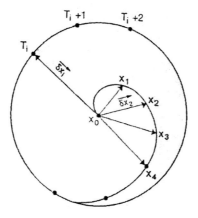

Figure 1.41. Poincaré section during a Hopf bifurcation. One passes from x_0, intersection of the limit cycle, to the circle of the successive points T_i, intersection of the T^2 toroid.

In the last case, the limit cycle (represented on the Poincaré section by the fixed point x_0) is transformed by the Hopf bifurcation in an attracting toroid T^2, whose Poincaré section is made from the points T_i on the a circle (provided $|\delta x_i|$ has a finite value). For this, the non-linear effects which had been neglected in the Floquet matrix transformation need to intervene to limit the increase of $|\delta x_i|$. Figure 1.41 (Bergé et al., 1984) shows the passing by instability from x_0, section of the limit cycle, to the toroid T^2 whose section is the circle of successive points T_i.

1.7.2. From the T^2 toroid to the T^3 toroid - Theory of Ruelle-Takens

Let us take the example of a viscous laminar flow. If the control parameter, here the Reynolds number, increases, the system loses its stability and starts oscillating with a period f_1. If the process is repeated, by increasing again the control parameter, a series of Hopf bifurcations may occur, with independent frequencies f_1, f_2, f_3.

The theory of Ruelle and Takens demonstrates that the T^3 toroid thus obtained can become unstable and be replaced by a strange attractor.

On the graph of the frequency spectra (Figure 1.42) and the system's evolution axis, in particular as a function of the control parameter, it is possible to see the transition from the periodic system to the quasi-periodic system with two frequencies, and then the transition to the chaotic system.

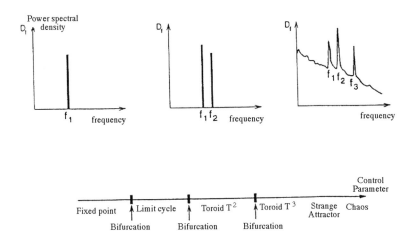

Figure 1.42. Theory of Ruelle and Takens. The successive transformations can be observed, from the T^2 toroid to the quasi-periodic T^3 toroid and to the chaotic state. The bottom diagram sums up the stages, B_1, B_2 and B_3 indicating the successive bifurcations.

The demonstration of Ruelle and Takens goes further, and shows that the T^3 toroid can be completely destroyed and replaced by a strange attractor, itself stable and not changed by new perturbations.

Contrary to the previous theory of Landau and Lifschitz (1971), Ruelle and Takens proved that a system with few degrees of freedom could engender a chaotic system. Even simpler models can be found to allow the transition to chaos, one of them being a T^2 toroid.

1.7.3 The Curry and Yorke Model

Instead of going to chaos by transformation of a T^3 toroid into a strange attractor, the model of Curry and Yorke (1977) goes directly from a quasi-periodicity with two frequencies to chaos. We will follow and sum up here the ideas developed and detailed by Curry and Yorke (1977) and Bergé et al. (1984).

This study was not rigorous, and was made numerically using the Poincaré section of a 3-D flow. To obtain chaos, the map must be non-linear, contracting and present a Hopf bifurcation.

The model results from the composition of two functions:

$$\mathcal{F} = \mathcal{F}_1 \, o \, \mathcal{F}_2 \tag{1.121}$$

where o means: followed by.

The map $\mathcal{F}_1$ is determined by the following relations between the polar coordinates of the points at iterations k and k+1 :

$$\mathcal{F}_1 : \begin{cases} \rho_{k+1} = \varepsilon \log.(1+\rho_k) \\ \theta_{k+1} = \theta_k + \theta_0 \end{cases} \tag{1.122}$$

$\theta_0 \geq 0$ is fixed; $\varepsilon \geq 1$ is the control parameter.

The next map $\mathcal{F}_2$ is using the linear coordinates:

$$\mathcal{F}_2 : \begin{cases} X_{k+1} = X_k \\ Y_{k+1} = Y_k + Y_k^2 \end{cases} \tag{1.123}$$

Note the similarity between $\mathcal{F}_2$ and the model of Hénon (see Section 1.5.5).

Figure 1.43 shows the results of the transformation $\rho_{k+1} = \varepsilon \log.(1+\rho_k)$ when ε is smaller or greater than 1. If $\varepsilon < 1$, ρ tends towards 0; if $\varepsilon > 1$, ρ tends towards a limit value ρ_L. Figure 1.43b shows the map $\mathcal{F}_1$; each iteration $\theta_{k+1} = \theta_k + \theta_0$ corresponds to a rotation of θ_0, and the result generally is the attracting circle C.

Two of the initial conditions were just met: dissipation and Hopf bifurcation. Indeed, when the control parameter passes through $\varepsilon = 1$, there is bifurcation of the limit cycle toward an invariant toroid. The last condition, non-linearity, is met by the map $\mathcal{F}_2: Y_{k+1} = Y_k + Y_k^2$, which breaks the decoupling of ρ and θ.

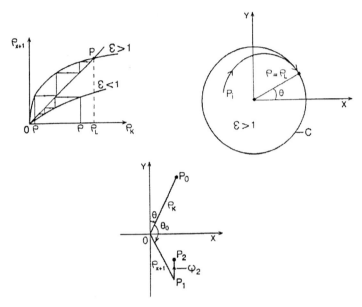

Figure 1.43. The Curry and Yorke model. The successive graphs show the two maps $\mathcal{F}_1$ and $\mathcal{F}_2$, their result for $\varepsilon > 1$, and the passage from P_0 to P_1 and from P_1 to P_2.

On Figure 1.43c, the map $\mathcal{F}$ allows one to go from P_0 to P_2, first by going from P_0 to P_1 (map $\mathcal{F}_1$), then by going from P_1 to P_2 (map $\mathcal{F}_2$). After a certain number of iterations (around a hundred), one sees in the plane a Poincaré section, which is the series of points representative of the attractor (or rather of its section xOy).

Figure 1.44 shows some examples for $\theta_0 = 2$ radians, and $\varepsilon = 1.27$ and then $\varepsilon = 1.48$. For $\varepsilon = 1.27$, the system is quasi-periodic with two frequencies f_1 and f_2; the section of the curve is a toroid $\mathcal{T}^2$.

Beyond a critical value $\varepsilon_c = 1.3953$, for example for $\varepsilon = 1.48$, bumps start appearing on the section, in infinite number and thickening it. The section's dimension becomes more than 1 (continuous line or circle) and is fractal. For $\varepsilon \geq \varepsilon_c$, the toroid is destroyed and the attractor becomes a strange attractor.

Ch. 1] Notion of Chaos 79

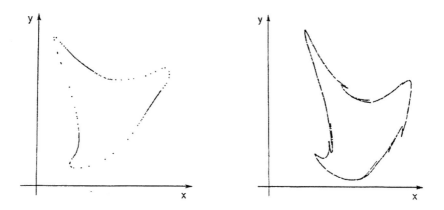

Figure 1.44. Results of the preceding map $\mathcal{F}$; (left) $\theta_0 = 2$ radians and $\varepsilon = 1.30$; (right) $\theta_0 = 2$ radians and $\varepsilon = 1.45$. From Curry and Yorke (1977).

It is easy to observe that close points on the section are quickly separated after a few iterations. This is the SIC, the property of strange attractors that was mentioned earlier. The points then get closer. Expansion and folding are conditions necessary and sufficient for chaos to exist.

The diagram of Figure 1.45 sums up the different stages of the path to chaos in the model of Curry and Yorke.

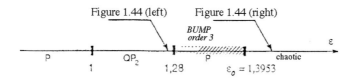

Figure 1.45. The path to chaos. Summary of the successive stages, in the model of Curry and Yorke, when ε increases.

1.7.4 Definitions of Chaos

As we could see all along in this chapter, there is no exact definition of chaos. Mathematically, Falconer (1990), acknowledges the existence of several possible definitions. He writes: "*if the map f has an attractor $\mathcal{F}$, then f has a chaotic behaviour on $\mathcal{F}$ or f may be considered as chaotic on $\mathcal{F}$, if the following conditions are fulfilled:*

(1) the trajectory $\{f^k(x)\}$ is dense on $\mathcal{F}$, for any $x \in \mathcal{F}$

(2) the periodic points of f on $\mathcal{F}$ (points such that $f^p(x) = x$, $p > 0$ integer) are dense on $\mathcal{F}$

(3) f exhibits SIC (Sensitivity to the Initial Conditions); i.e. there is a number $\delta > 0$ such that, for any $x \in \mathcal{F}$, there are points $y \in \mathcal{F}$, arbitrarily close to x which verify $|f^k(x) - f^k(y)| \geq \delta$, for any k. Thus points initially close to each other will not stay close after iterations of f"

We used for dynamic systems the term of determinist chaos, by opposition to the term of chaotic chaos. This is to separate systems[5] when the dimension of the phase space increases, contrary to other systems with a high number of degrees of freedom.

A pseudo-definition of chaos by Bergé et al. (1984) considers chaos as a characteristic loss of memory of a signal about itself, which makes unpredictability a dominant feature of its definition.

The Control of Chaos

The Sensitivity to Initial Conditions enables the control of a chaotic system, which would not be possible for a stable system. A very slight intervention on a chaotic system will have important consequences, which is not the case for a non-chaotic stable system. An example of control is given in Annex A.3. The consequences of these new approaches are extremely promising.

This method is known as OGY, from the names of its creators: Ott, Grebogi and Yorke. It consists in studying the chaotic system in its phase space through the analysis of a Poincaré section (Ott et al., 1990). The system is left to itself on this section until it approaches the periodic orbit chosen, and by changes of the appropriate parameter, the system is forcibly maintained on this orbit. The method's advantage is that it does not require a detailed model of the chaotic system, but only a little information about the Poincaré section selected (see Annex A.3).

---

[5] some of which fulfil the conditions above; SIC, saturation of the Grassberger exponent and Procaccia's correlation function.

2

Applications to Geophysics

> *Pure the Earth and pure the Light*
> *Pure were they all,*
> *Each being what it was*
> *Each doing faultlessly and wonderfully,*
> *What it was formed to do*
> Paul Valéry, Paraboles, 1935

As mentioned in the introduction, the application of non-linear dynamics to geophysics first concerns the study of the Earth's morphology at all scales, from topography to geomorphology and altimetry. Historically, this is the domain where the first applications of the works on fractal sets were developed (Richardson, 1961; Mandelbrot, 1967). Rock fragmentation and fracturation form the link between rheology and geosciences. We saw that numerous studies were accomplished in the physics of solids (de Gennes, 1976; Guyon, 1978; Roux, 1990). We shall not delve into their most arcane details, but we will use their results in geological areas, such as fracturation, fault systems, and percolation. This will lead us to seismology and volcanology, choice domains for the non-linear approach. The introduction of the time parameter will lead from fractal geometry models to unstable dynamic systems, with their probabilistic implications and the introduction of Self-Organised Criticality (SOC), which may have been found in the theoretical section of this book, but will be seen here only through its seismologic and volcanologic applications. Finally, the many applications to geomagnetism seem specific enough to geophysics, because of their direct consequences for the study of mass transfer inside the terrestrial core. The applications of non-linear dynamics that we present here are voluntarily limited in number and detail, and in the Annexes, the curious reader will find some other developments which will show the wide range of possible applications.

2.1 GEOMORPHOLOGY

The first applications of fractal geometry dealt with geomorphology. Beautiful relief images were generated during the first works of Mandelbrot (1965, 1984, 1989), for example his famous sunset over a fractal island. We also saw the importance of the study of the length of rocky coasts by Richardson (1961) in the definition of fractal dimension put forward by Mandelbrot (1967).

2.1.1 Observations

As they appear on continents or on the seafloor, terrestrial reliefs are complex shapes which resist classical analysis. Among the purest ones are the submarine reliefs close to mid-ocean ridges and spreading centres. Away from mid-ocean ridges, the shapes are smoothed by sediment deposition. And continental reliefs result from the interaction of several factors, meteorologic erosion being the chief. Colour Plate 2 (between pp. 192 and 193) shows the submarine relief on the Mid-Atlantic Ridge close to the Azores, as revealed by multibeam bathymetry.

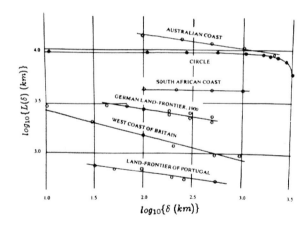

Figure 2.1 Approximate coast lengths, from Richardson (1939). In log-log coordinates, the slope of the line representing the function (coast length; measuring interval) is 1-D (from Mandelbrot, 1975).

Coming back to the length of Brittany's coasts, and applying the definition of Section 1.3, $N_i = C/r_i^D$, where N_i is the number of steps of length r_i, the perimeter is given by: $P_i = N_i r_i = C r_i^{1-D}$ (Figure 2.1). Mandelbrot (1967) demonstrated that the length of Brittany's coasts satisfies this equation with a dimension $D = 1.25$ (see Figure 2 in the Introduction). This method can be used to determine the fractal dimension of a relief from the length of its perimeter on a contour line, as a function of the measuring interval. In mountain zones, there generally is a good agreement between this equation and $1.3 < D < 1.5$ (Turcotte, 1989). D is weakly correlated with the tectonic style or the age of rocks, but there are exceptions to this law, such as young volcanic islands (Goodchild, 1980). An unpublished personal work on recent seamounts of the Tehaitia region in French Polynesia shows that these dimensions vary from one contour line to the next.

The application of this method to hydrographic networks makes it possible to establish their fractal nature and the relation between fractal dimension and several natural parameters (Beauvais et al., 1994). An example of the application of Richardson's method to the morphometric study of rivers in Central Africa is provided in Annex B.1.

When building fractal sets, in geomorphology, the profiles used often possess some vertical exaggeration. This introduces the definition of **self-affine sets**. These sets, including in particular the self-similar sets, result from an **affine transform** S:

$$S: \mathcal{R}^n \to \mathcal{R}^n; \; S(\mathbf{x}) = T \cdot \mathbf{x} + \mathbf{b} \qquad (2.1)$$

where T is a linear transform on $\mathcal{R}^n$ which can be represented by an n-n matrix, **x** and **b** are vectors of $\mathcal{R}^n$ (**x** represents the object to transform, **b** is given with T to define the transform; see Section 4.1).

An affine transform S can therefore be a combination of translation, rotation, dilation, contraction or even reflexion. Unlike similarity transforms, the affine transforms contract or dilate with different ratios for different directions.

2.1.2 Fractal Self-Affine Topographic Profiles

The studies of topography imply two coordinates leading to self-similarity, as the two horizontal coordinates are in fact indistinguishable. However, the altitude h is not self-similar to the two horizontal coordinates, but, taken independently along a linear path, it is an example of self-affine fractal. The following developments are borrowed from Huang and Turcotte (1989).

For a self-affine fractal, the spectral energy density S for the profile (obtained from the Fourier transform of a topographic profile - see Malinverno and Gilbert, 1989) must have a power-law dependency on the wavenumber k (Mandelbrot, 1986):

$$S(k) \approx k^{-\beta} \qquad (2.2)$$

The power-law dependency of the spectra of topographic profiles h(x) (with k being the Fourier transform of x) can be related to the fractal dimension by using the variance V, measuring the topography variations:

$$V(L) = \frac{1}{L} \int_0^L [h(x) - \bar{h}]^2 \, dx \qquad (2.3)$$

where L is the profile's length and h̄ the mean height on L.

$$\bar{h}(L) = \frac{1}{L} \int_0^L h(x) \, dx$$

For the profile to be fractal, the variance V(L) must necessarily be a power-law (Voss, 1985 a,b):

$$V(L) \sim L^{2H} \tag{2.4}$$

The standard deviation σ is linked to the profile length L by:

$$\sigma(L) = [V(L)]^{1/2} \sim L^H \tag{2.5}$$

For a fractionary Brownian movement (cf. Mandelbrot, 1982), $H \in \,]0\,;1\,[$. To measure this fractal dimension, a reference box is chosen with a length L and height σ (Voss, 1985). If the fractal object had been self-similar, the box would have been a square, but rectangular boxes must be chosen for self-affine fractals.

Let us take a series of boxes of the n-th order, of width $L_n = L/n$ and of height $h_n = \sigma/n$, n integer. The number N_n of boxes of the n-th order necessary to cover a length L and a height $\sigma_n = \sigma(L/n)$ is:

$$N_n = \frac{L \, \sigma_n}{L_n \, h_n} = \frac{\sigma_n}{\sigma/n^2} = n^2 \frac{\sigma_n}{\sigma} \tag{2.6}$$

According to equation (2.5), $\sigma(L) = L^H$ and $\sigma(L/n) = (L/n)^H$. Equations (2.5) and (2.6) therefore lead to:

$$N_n = n^2 \frac{(L/n)^H}{L^H} = n^{2-H} = \left(\frac{L}{Ln}\right)^{2-H} \tag{2.7}$$

Comparing equation (2.7) with the definition of fractal distribution, we get:

$$D = 2 - H \tag{2.8}$$

The energy spectral density S(L) has a power-law dependency on the length L and can be related to the variance V(L) with:

$$S(L) = L \, V(L) \approx L^\beta \approx L^{1+2H} \tag{2.9}$$

$\beta = 1 + 2H$, and, using equation (2.8), we get:

$$D = (5 - \beta) / 2 \tag{2.10}$$

On a one-dimensional profile, D is supposed to range between 1 and 2, and the corresponding value of β therefore satisfies: $1 < \beta < 3$.

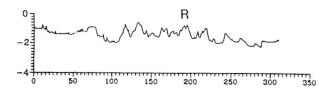

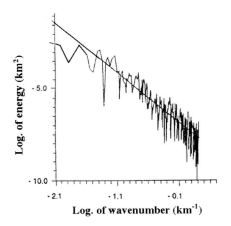

Figure 2.2. Spectrum of a bathymetric profile south of the Azores: (top) profile across the axis of the spreading centre, shown by the letter R, with distances in kilometres; (bottom) log-log graph of the spectral energy as a function of the wavenumber, with a least-squares linear regression. From Ballu, 1992.

For Brownian noise, $\beta = 2$, $H = 0.5$ and $D = 1.5$. Many one-dimensional studies of bathymetry and topography have been performed with Fourier transforms (e.g. Cox and Sandstrom, 1962; Bretherton, 1969; Warren, 1973; Bell, 1975, 1978; Barenbatt et al., 1984; Fox and Hayes, 1985; Gibert and Courtillot, 1987). These studies concluded to a good correlation with $S(k) \sim k^{-\beta}$, $\beta \approx 2$ ($D \sim 1.5$) for wavelengths between 10^3 and 10^{-1} km; topography therefore looks like a Brownian noise.

2.1.3 2-D Representation

Examples of 2-D representations are topographic maps with contour lines or bathymetric maps obtained with multibeam sonars.

In an NxN-grid, the N^2 values $h_{n,m}$ (n, m integers) can undergo a Fourier transform. The NxN-array of the complex coefficients $H_{s,t}$ is obtained with:

$$H_{s,t} = \sum_{n=0}^{N-1} \sum_{m=0}^{N-1} h_{n,m} \exp\left[-\frac{2\pi i}{N}(s_n + t_m)\right] \qquad (2.11)$$

where s is the transform in the x-direction (s = 0, 1, 2, ..., N-1), and t is the transform in the y-direction (t = 0, 1, 2, ..., N-1). Each coefficient $H_{s,t}$ is associated to an equivalent radial number $r = (s^2 + t^2)^{1/2}$.

The mean spectral energy density S_j is given for each radial number k_j by:

$$S_j = \frac{L}{N_l} \sum_{1}^{N_l} |H_{st}|^2 \qquad (2.12)$$

where N_l is the number of coefficients that satisfy $j < r < j+1$. If the spectral energy density S_j has a power-law dependency on k_j, the fractal dimension D can be deduced from the slope β of this power-law (Voss, 1985):

$$D = (7 - \beta)/2 \qquad (2.13)$$

This is the extension to two dimensions of the law for one dimension: $D = (5 - \beta)/2$.

Many authors have applied the preceding to some real examples. Sayles and Thomas (1978) and Berry and Hannay (1978) were the first to discuss the importance of studying the slope to better describe the energy spectra in the log-log space. Fox and Hayes (1985) and Gibert and Courtillot (1987) did not use the vocabulary of fractal geometry, but computed the slope and starting ordinate of regression lines plotted for topographic or bathymetric (or altimetric) spectra, in order to quantify the topography roughness.

Brown and Scholz (1985) computed the energy spectra and the fractal dimension D to describe the surface of rocks. They observe that the regression lines for these spectra can be divided into several lines with different slopes, and that D may vary with spatial scales. Gilbert (1989) studied a bathymetric profile in the South Atlantic, starting from the axis of the Mid-Atlantic Ridge and ending at anomaly 31. Residual depths are obtained by subtracting from bathymetry the value from the depth-age relation of Parsons and Sclater (1977). This assumes that the spreading rate has been constant through time. As the zone studied on the seafloor is bigger than the sampling step (200 m compared to 66.66 m), he combines three different resampling techniques to smooth the profiles. The spectrum is obtained by plotting the FFT on a log-log diagram and computing the regression line (Figure 2.3). For the Sierra Nevada profile, the author used a Digital Terrain Model (DTM) with a UTM grid sampled every 30 metres.

The values of D, i.e. the fractal dimension of bathymetric and topographic profiles, range between 1.12 and 1.32 in the Atlantic (depending on the technique used), between wavelengths of 1 and 10 km. They range between 1.21 and 1.47 for the same wavelengths in the Sierra Nevada. The author concludes that topography and bathymetry satisfy to fractal geometry between 1 and 10 km, but that their quantification must obey two points: (1) the scale must be specified with the length range used; (2) the analysis techniques must be clearly indicated as they influence the results (for the same technique, D can be significantly different).

Annex B.1 presents a study (Panteleiev et al., 1994) using the method of Dubuc et al. (1989) to analyse fractally the altimetric geoid on the oceans.

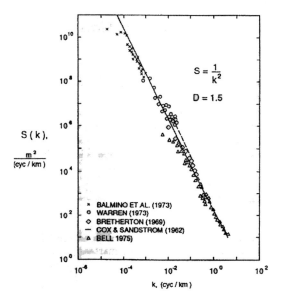

Figure 2.3. Spectral density spectrum. Spectra of terrestrial topography have been compiled and shown on this log-log graph as a function of the wavenumber (Turcotte, 1989).

2.2 FRAGMENTATION, FRACTURATION, TECTONICS, AND PERCOLATION

The first introduction to these problems was made in an article by Allègre et al. (1982), where the fractal behaviour of rock fracturation was precisely described, along with the self-similarity and hierarchy in fracturation models. This approach was followed by Smalley et al. (1985) and Turcotte (1986). It resulted in a relatively simple formulation of these seemingly complex problems, which we shall present here.

2.2.1 Fragmentation

<u>The observables</u>

When a rock is fragmented (for example with a hammer, as for breaking up a lump of sugar), the sizes of fragments vary considerably, from dust to portions nearly as big as the original sample.

The fragmentation mechanism may be related to the distribution of joints and the pre-existence of weakness planes inside the rock. This will be assumed to be the case when looking at fragmentation models.

Many statistical relations have been established to correlate the distributions of fragment sizes. Distributions can be divided into two categories, lognormal and power-law. Turcotte (1986) showed that the power-law distribution is equivalent to a fractal distribution. Table 2.1 shows series of values obtained when fragmenting coal blocks (Bennett, 1936) or gabbros, when looking at fragments created by meteorite impacts, or at the fragmentation of a granite diapir in a nuclear explosion (Schoutens, 1979).

Fragment distributions

The distribution law of fragments can be studied as a size-frequency relation. It may then be possible to recognise whether this distribution is exponential or Poissonian, whether it is a power-law or a Weibull distribution.

If $N(m)$ is the number of fragments of mass greater than m, $N(m) = c\, m^{-b}$, with c constant. And remarking that $m \sim r^3$, we get: $N(m) = c\, r^{-3b}$.

This corresponds to the definition of a fractal set $N \sim r^{-D}$ (see Section 2.3) with $D = 3b$. This power-law distribution is therefore equivalent to a fractal distribution.

A Weibull-type distribution may sometimes be observed experimentally. It is expressed as:

$$\frac{M(r)}{M_T} = 1 - \exp\left(-\left(\frac{r}{\sigma}\right)^\alpha\right) \tag{2.14}$$

whre $M(r)$ is the cumulated mass of the fragments of radius smaller than r (and the radius is (volume)$^{1/3}$), M_T is the total mass and σ the mean size of fragments.

For $r/\sigma \ll 1$, this comes back to a power law:

$$\frac{M(r)}{M_T} = \left(\frac{r}{\sigma}\right)^\alpha \tag{2.15}$$

Differentiating, we get:

$$dM \approx r^{\alpha-1}\, dr \tag{2.16}$$

Using the definition $N \sim r^{-D}$ and differentiating, $dN \sim r^{-D-1}\, dr$. And, as $dN \sim r^{-3}\, dM$, these three equations can be combined into:

$$r^{-D-1} \sim r^{-3}\, r^{\alpha-1} \tag{2.17}$$

That is: $D = 3 - \alpha$.

This Weibull-type distribution is equivalent, for $r/\sigma \ll 1$, to a fractal distribution.

Table 2.1 Fractal dimensions observed when fragmenting miscellaneous objects (Turcotte, 1989).

Objects	Fractal dimension D
Gabbro fragmented by a lead projectile	1.44
Idem with a steel projectile (Lange et al., 1984)	1.71
Meteorite fragments (McCrosky, 1968)	1.86
Weakness plane models (Turcotte, 1986)	1.97
Disaggregated gneiss (Hartmann, 1969)	2.13
Disaggregated granite (Hartmann, 1969)	2.22
Fragmentation by a chemical 0.2 kT explosion (Schoutens, 1979)	2.42
Fragmentation by a nuclear 62 kT explosion (Schoutens, 1979)	2.50
Coal fragments (Bennett, 1936)	2.50
Interstellar dust (Mathis, 1979)	2.50
Fragmented basalt (Fujiwara, 1977)	2.56
Clay sand (Hartmann, 1969)	2.61
Gravels and sands from alluvial terraces (Hartmann, 1969)	2.82
Fragile pillar model (Allègre et al., 1982)	2.84
Glacial alluvia (Hartmann, 1969)	2.88
Rocky meteorites (Hawkins, 1960)	3.00

Finally, let us remark that the exponential distribution is very distinct from the fractal distribution, as it is linear on a semilog graph instead of a log-log graph. This difference will prove to be very promising for probabilities (see Section 3.4): the former distribution will be random, the others will be determinist.

Models

To detail the possible mechanisms, we will follow the approach of Allègre et al. (1982) and Turcotte (1986). It uses renormalisation groups (Madden, 1976, 1983). The basic hypothesis is that the probability p_c that a cell (e.g. a cube in the simplest case) gets fragmented into 8 elements is the same whatever the size level considered. If the initial object has a size h, the 8 largest fragments have a size h/2 (Figure 2.4), and they themselves are divided into 8 fragments h/4, etc.

At the m-th fragmentation stage, the total number of elements is:

$$N_m = (1 - p_c) \left[1 + 8\, p_c + 8\, p_c^2 + \ldots + 8 p_c^m \right] \quad (2.18)$$

From the relationship $N \sim r^{-D}$, we get:

$$\frac{N_{m+1}}{N_m} = \frac{h^{-D}/2^{-D}}{h^{-D}} = 2^D \quad (2.19)$$

This ratio is a function of p_c:

$$\frac{N_{m+1}}{N_m} = 8\, p_c \quad (2.20)$$

And therefore:

$$8\, p_c = 2^D \text{ and } D = \frac{\log . p_c}{\log . 2} \quad (2.21)$$

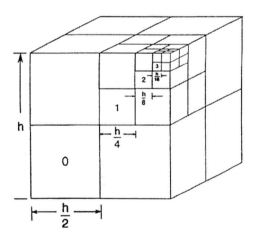

Figure 2.4. Cube fragmentation model. A cube of side h can be fragmented in eight cubes of size h/2, which can themselves be fragmented in eight cubes of size h/4, etc. From Allègre et al. (1982), and Turcotte et al. (1986).

We know that $1/8 < p_c < 1$, and therefore D will vary between 0 (for $p_c = 1/8$) and 3 (for $p_c = 1$). Let us remember that, from one level to the next, p_c is the probability for a cube to get fragmented into 8 elements and that, once defined, it remains the same at all levels.

Figure 2.5 shows the variation of D as a function of p_c. It is possible to go further into the model to compute the relation between the probability p_n that a cell of the n-th order is fragile and the probability p_{n+1} that an n-th order element (i.e. a cell of the (n+1)-th order) is fragile. The characteristic dimension of an n-th order cell is $(h/2)^n$ and the the dimension of an (n+1)-th order element is $(h/2)^{n+1}$. This relation was established by Allègre et al. (1982) in the model of the "sound pillar" and by Turcotte (1986) for the model of the "weakness plane".

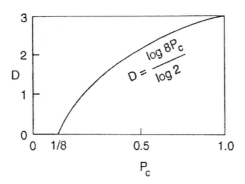

Figure 2.5. Fragmentation probability and dimension. D varies as a function of p_c, fragmentation probability ranging between 1/8 and 1 (Turcotte, 1986).

Model of the "sound pillar"

The weakness of a particular cell (cube of n-th order) is determined by the weakness of its elements (cubes of (n+1)-th order). In each cell, there may be from 0 to 8 fragile elements, which leads to $2^8 = 256$ possible combinations which can be reduced into 22 different topological configurations (Figure 2.6). These cells will be considered as "sound" or solid if they contain a sound pillar (i.e. a cornerstone) made from two solid elements (Figure 2.7). In the other cases, the cell will be fragile. This configuration is verified in 6 configurations out of 22 (Figure 2.6). The following relationship can be deduced (Turcotte, 1986):

$$p_n = p_{n+1}^8 + 8\, p_{n+1}^7 (1-p_{n+1}) + 16\, p_{n+1}^6 (1-p_{n+1})^2 + 8\, p_{n+1}^5 (1-p_{n+1})^3 + 2\, p_{n+1}^4 (1-p_{n+1})^4 \qquad (2.22)$$

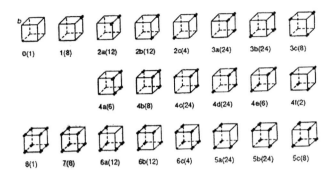

Figure 2.6. Possible combinations during the fragmentation of a "sound pillar". There are 256 possible combinations, which reduce to 22 different topological configurations (Allègre et al., 1982).

The graph of the relation expressed in equation (2.22) has been plotted in Figure 2.8, along with the bisecting line $p_n = p_{n+1}$. This is a typical first-return map (see Section 1.5.7). The points 0 and 1 are stable, fixed points. The intersection of the curve with the bisecting line occurs at $p_n = p_{n+1} = p_c = 0.896$, point separating the two regions $p > p_c$ and $p < p_c$. p_c is the critical probability, corresponding to a catastrophic fragmentation of the object. Plotting the next iterations of the renormalisation group show that, for values of $p_{n+1} < p_c$, one tends toward the point 0; there is decreasing fragility. Starting, from example, from $p_{n+1} = 0.6$, one finds $p_n = 0.2723$, $p_{n-1} = 0.0118$, $p_{n-2} = 3.9 \cdot 10^{-8}$, etc. The object remains solid. Conversely, for values of $p_{n+1} > p_c$, one tends toward the point 1; there is increasing fragility. The object gets fragmented as n decreases. There is a bifurcation for $p_{n+1} = p_c$. The fractal value associated to this catastrophic fragmentation can therefore be computed, and is $D = 2.84$.

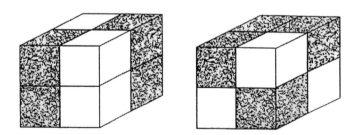

Figure 2.7. Model of the "sound pillar". The initial cell is solid if it contains a sound pillar (i.e. a cornerstone) made from two sound elements. The left cube is an example of a "sound pillar", whereas the right cube is fragile (Allègre et al., 1982).

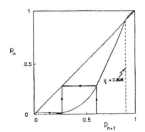

Figure 2.8. Graph (p_n, p_{n+1}) in the "sound pillar" model. In this first-return map, the intersection with the first bisecting line gives $p_n = p_{n+1} = p_c = 0.896$, where p_c is the critical probability corresponding to a catastrophical fragmentation of the object.

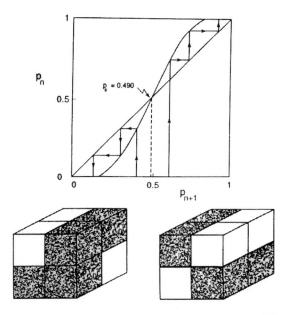

Figure 2.9. Model of the "weakness plane" (Turcotte, 1986). The first-return map (p_n, p_{n+1}), where $p_c = 0.490$, is plotted above.

Model of the "weakness plane"

The model of Figure 2.6, with its 22 topological configurations, is used as a starting point. The cells are now considered fragile only when the fragile elements make up a plane crossing the cell (Turcotte, 1986).

9 configurations out of 22 correspond to this situation. The relation between p_n and p_{n+1} now becomes:

$$p_n = p_{n+1}^8 + 8 p_{n+1}^7 (1-p_{n+1}) + 28 p_{n+1}^6 (1-p_{n+1})^2 + 56 p_{n+1}^5 (1-p_{n+1})^3$$
$$+ 38 p_{n+1}^4 (1-p_{n+1})^4$$

$$\Leftrightarrow p_n = 3 p_{n+1}^8 - 32 p_{n+1}^7 + 88 p_{n+1}^6 - 96 p_{n+1}^5 + 38 p_{n+1}^4 \qquad (2.23)$$

Plotting the graph of the first-return map (p_n, p_{n+1}) shows that $p_c = 0.490$, which corresponds to a catastrophic fragmentation of dimension $D = 1.971$.

These models are correctly accounting for the fragmentation processes which lead to the power-law fragment distribution observed in the experiments. Furthermore, models of this kind are internally consistent, as they verify the starting hypothesis that breaks can occur at joints or along weakness planes. Our choice of a cube as initial cell was made for an easier understanding, but it should be clear that other types of geometries could have been chosen (e.g. crystal lattices in crystallography and mineralogy).

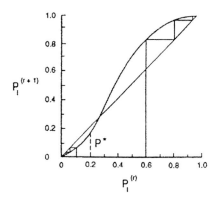

Figure 2.10. Two-bump model. The critical probability is $p_c = 0.2063$ (Smalley et al., 1987).

Other examples

It is of course possible to choose in the models a finer fragmentation, with each cube of size h dividing at each stage into 64 cubic elements of dimension h/4. Smalley et al. (185) used this renormalisation approach to investigate the behaviour of faults, supposedly constrained at all scales by the distribution of bumps and boundaries (Aki, 1981). The authors assume that the bumps and boundaries have a statistical distribution of resistance, and they introduce a mechanism of transfer of constraints to neighbouring bumps when one bump is broken.

In a statistical approach close to the previous one, but computationally heavier, they establish a relation between p_n and p_{n+1}, for cells with one or two bumps, with a quadratic Weibull distribution of their resistance. They can determine a critical breaking probability $p_c = 0.2063$ for two bumps and $p_c = 0.1707$ for 4 bumps. For $p_1 < p_c$, the successive iterations lead to $p_1^\infty = 0$ and there are no breaks. Conversely, for $p_1 > p_c$, the successive iterations lead to $p_1^\infty = 1$ and the system breaks apart; i.e. the fault is forming (Figure 2.10).

2.2.2 Rock Fracturation

Theoretical studies and techniques have been largely used in rock fracturation. Analysis of the geometry of fracture fields shows that, despite their random aspect, it is possible to bring into evidence a statistical internal invariance, when changing the study scale. This stems of course from the fragmentation laws studied earlier. It is therefore logical to expect fracture fields to be amenable to fractal analyses.

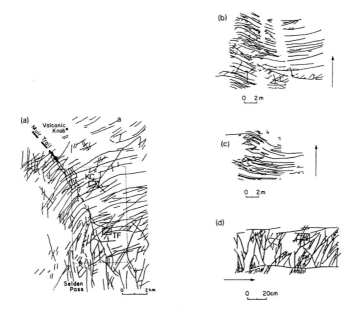

Figure 2.11. Self-similarity in fracturation. Four images of fracture fields observed at different scales (Velde et al., 1991).

<u>General observation of fault and fracture fields</u>

Before developing techniques for the analysis of fault and fracture fields, let us observe them at different scales. As remarked by Allègre et al. (1982), the geologist can look at fractures on thin slices observed with a microscope, on the rock outcrop itself, or on satellite images. The confusion between images is easy, as visible in Figure 2.11. This brings back to mind the self-similarity of fractal sets, and this property will indeed be demonstrated for a granite dome from Montana.

Measuring the length of fractures or faults

This method has been exposed in several articles (Sornette et al., 1990; Davy et al., 1990). It consists in plotting a circle of centre M and radius r, at the centre of a network of fractures or faults, on a plane crossing the rock where the trace of fractures is highly visible. This can be done on a photograph or a map of the surface. Inside this circle, there are a number of faults, or fault segments, with lengths $l_1, l_2, ..., l_n$. These values are summed for the complete circle of radius r, giving $L(r) = l_1 + l_2 + ... + l_n$. r is then increased, giving a new value of $L(r)$ (Figure 2.12). Generally, $L(r)$ increases as a power of r; $L(r) \approx r^D$. The exponent D is the fractal dimension of the network.

This is achieved practically by plotting log.L(r) and log.r (Figure 2.12). If the relation between the two is a power law, the different points are aligned. The regression line has a slope D. From what was said earlier about fractal sets, we know that the dimension D is such that: $1 < D < 2$. Davy et al. (1990) looked at an experimental model of the India-Asia collision and found a value of 1.73 ± 0.05.

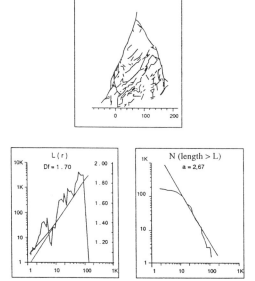

Figure 2.12. Fractal analysis of a fault field. The fractal dimension of a fault network measured on a plan (tectonic map, above) is obtained by measuring the length l_1 of the fault network in a circle of radius r_1, its length l_2 in a circle of radius r_2, ..., and looking at the slope D of the resulting log-log graph. From Sornette et al. (1991).

One of the main questions allowed by this approach is the preservation of vast undeformed regions in the middle of heavily fractured areas. Several authors had interpreted these observations as resulting from the stronger resistance of these zones. Fractal analysis shows that the existence of undeformed zones may naturally come from the fractal character of continental tectonics, and not necessarily be associated to lateral heterogeneity of the continental lithosphere.

Cantor dust method

This specific method will be found again in Section 3.3, applied to time series of events such as volcanic eruptions or seismic crises. It is applied here to the spatial domain. The method is exposed in detail in the article by Velde et al. (1990). It consists in crossing the fracture field with a line whose direction relative to the fracture field is noted. The intersection points of the line and the fractures produce a series of points, studied through a box-counting method (Ledésert et al., 1993, a, b).

The study interval is covered by segments of length u_i, covering one or several events (here, the events are the intersection points). The number of segments is $N_i = 1/u_i^D$ if the distribution of points is fractal. In the case of the triadic Cantor set (see Section 1.3.2), for two successive iterations i and i+1, we have: $N_{i+1}/N_i = 2$ and $u_{i+1}/u_i = 3$, and therefore:

$$D = - \frac{\log.(N_{i+1}/N_i)}{\log.(u_{i+1}/u_i)} = \frac{\log.2}{\log.1/3} = 0.6309 \qquad (2.24)$$

The parameter generally considered is the fraction x_i of segments (or steps) with length u_i which contain dust. The relation between x_i and u_i is:

$$N_i u_i = x_i L \; ; \; N_{i+1} u_{i+1} = x_{i+1} L \qquad (2.25)$$

L is the length of the study interval. Therefore:

$$\frac{x_{i+1}}{x_i} = \frac{N_{i+1} u_{i+1}}{N_i u_i} = \frac{u_{i+1}}{u_i}^{1-D} \qquad (2.26)$$

In the case of the triadic Cantor set;

$$\frac{x_{i+1}}{x_i} = \frac{u_{i+1}}{u_i}^{1-D} = \frac{2}{3} \qquad (2.27)$$

And therefore:

$$D = \frac{\log.2}{\log 3} = 0.6309 \qquad (2.28)$$

If the fragmentation becomes more general, by varying the ratios N_{i+1}/N_i and u_{i+1}/u_i, D varies between 0 and 1. If there is only one point, $D = 0$, which is the Euclidean dimension of a point. And when all the segments are filled, $D = 1$, which is the Euclidean dimension of a segment.

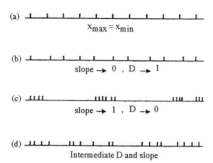

Figure 2.13. Application of the Cantor dust method to the fractal analysis of a 2-D section from a fracture field. The intersection of lines determines a series of points, which is processed by the method of Smalley et al. (1987): (a) the slope is close to 0 when the events are evenly spaced; (b) the slope is still close to 0 when the spacing between events does not vary much; (c) the slope tends toward 1 (and D tends toward 0) when the events are clustered; (d) and in the intermediate cases, the slope and the dimension D have values ranging between 0 and 1. From Velde et al., 1990.

One of the main problems encountered in this type of studies is related to the number of events, often too small to guarantee a significant statistical analysis. To increase this number, it is possible to use a series of parallel lines covering the surface to study (Velde et al., 1990, 1991), and the successive lines are read as the series of lines in a book (Figure 2.13). The intervals between lines must be large enough that the results are not repeated too often from one line to the next.

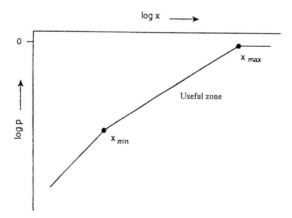

Figure 2.14. Cantor dust method. The log-log graph is the result of the analysis performed in Figure 2.13. The useful zone $X_{min} < X < X_{max}$ enables the measuring of the slope m and the fractal dimension $D = 1 - m$.

This choice of an average spacing u_i has been studied empirically, but would benefit from a theoretical analysis. To ease the reading of successive series of points, we have personally developed an automatic analysis system based on a camera looking at photographs of the sections.

The final log-log graph of this analysis shows up like a line broken into three segments (Figure 2.14). The first segment has a slope of 1 (i.e. a dimension of 0) for values of $0 < u < u_{min}$. It corresponds to small values of x_i for which x_i/x_{i+1} is close to 1. The slope of log.x as a function of log.u is close to 1. This part of the graph is not significant.

For $u_{min} < u < u_{max}$, we are in the useful zone where $D = 1 - m$, m being the segment slope.

u_{max} is the value of u_i where all u_i cover some dust. Therefore $x = 1$ and $m = 0$ (because log.x = 0), and $D = 1$. This means that, for this value of u, the segment is filled.

For the fractal character of the series to be recognised, the useful zone is generally defined to cover at least one order of magnitude (Davy et al., 1990), i.e. $u_{max}/u_{min} \geq 10$. This last condition of self-similarity is largely verified on the example of a granite done in the Sierra Nevada, where the value of D is constant at scales ranging from the scale of satellite imagery down to the scale of thin slices, i.e. 5 orders of magnitude (Velde et al., 1991).

The method can be applied for different angles of the intersecting line, and variations of D are observed. They correspond to anisotropy in the fracture field. It is tempting to plot the variations of D with the angles, as is done when studying the anisotropy of other physical parameters. But the interpretation is not straightforward. Harris et al. (1991) and Velde and Dubois (1991) showed that only two values of D could be "seen". This is coming from the properties of projections and intersections of fractal sets exposed in Section 1.3.8.

Indeed, if the line $E(\theta)$ is infinite on an infinite plane, and if the perpendicular sets F_1 and F_2 of lines, distributed fractally with the respective dimensions D_1 and D_2, are as large as possible, the points of intersection on $E(\theta)$ are distributed along two Cantor sets with the respective dimensions D_1 and D_2 (Figure 2.15). Because of the property mentioned earlier, the dimension of the resulting set is :

$$\max.\{\dim(E(\theta) \cup F_1), \dim(E(\theta)) \cup F_2)\} \qquad (2.29)$$

For the example shown in Figure 2.15, $D_1 = 0.25$ and $D_2 = 0.75$. The dimension of the resulting set is therefore 0.75.

When $E(\theta)$ is infinite, this is true for any value of θ, except when θ is such that $E(\theta)$ is parallel to F_2 (i.e. perpendicular to F_1), and then the dimension suddenly becomes 0.25.

It appears therefore that only two subsets (if any) can be observed. The first subset has the higher dimension, the second subset only appears when $E(\theta)$ is parallel to the fractures of the first set. The sets with dimensions smaller than these two values would be "invisible".

In reality, the fracture field $\mathcal{F}_1$ does not usually consist in parallel fractures, but shows a statistically privileged direction. There will therefore be some noise when the angle θ will be in this direction ($\mathcal{E}(\theta)$ parallel to this direction).

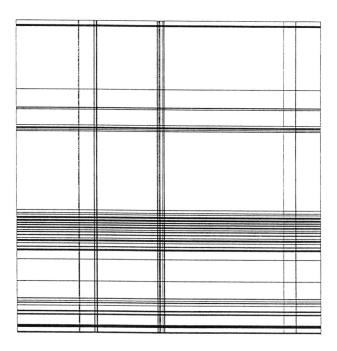

Figure 2.15. Intersection of two sets $\mathcal{F}_1$ and $\mathcal{F}_2$. They are made from fracture fields with perpendicular directions, $D_1 = 0.25$ and $D_2 = 0.75$. In general, the dimension of the resulting set $\mathcal{E}(\theta)$ along the angle θ is expressed as max.$\{\dim(\mathcal{E}(\theta) \cup \mathcal{F}_1), \dim(\mathcal{E}(\theta)) \cup \mathcal{F}_2)\} = 0.75$. The exception is when the intersecting line is parallel to $\mathcal{F}_2$, and the dimension jumps to 0.25. From Harris et al., 1991.

In summary, two factors can perturb the law established earlier:

1. The fractures of a given set are not completely parallel.
2. The plane of study and the fracture fields are not infinite. This limits the number of intersection points, especially when θ gets close to the critical value (parallel to the direction of the fractures in the set with the highest value of D). This goes back to the situation of short series evoked earlier.

In the examples mentioned above, we could as well have applied the method developed by Grassberger and Procaccia (1983), using the fact that the intervals of "sound" rocks between successive fractures, mapped along the line intersecting the fracture fields, correspond to a series of discrete values. This series can be tested with the correlation function, which gives the dimension of the attractor (in the spatial domain, and not in the time domain as is commonly done for dynamic systems). This method was successfully applied the position of fractures in drilling data (Dubois et al., 1993).

2.2.3 Tectonics - Study of Surface Faults

Because of self-similarity, it is possible to apply the techniques outlined in the previous chapter to aerial photographs, satellite images or tectonic maps. This was for example accomplished by Velde et al. (1990), who applied the Cantor's dust technique to the granite dome of Mt. Abbot in the Sierra Nevada. Another example is the study by Davy et al. (1990) of the fault lengths in the Indian-Eurasian collision.

The many box-counting techniques proposed by Falconer (1990) (see Section 1.3) have been applied in two articles about the San Andreas fault system (Aviles et al., 1987; Okubo and Aki, 1987).

The first article (Aviles et al., 1987) uses the technique outlined in Section 2.1, where broken lines (coasts, rivers, contour levels) are measured with segments of variable length between 0.5 km and 1000 km (Figure 2.16). The relatively small values of D are physically interpreted as showing the irregularities of the fault surface expression. After observing variations along the main fault, the authors identified six different segments, corresponding to very different seismic regimes. The dimensions observed for these segments vary from 1.0008 to 1.0191. The small values, close to 1, show the generally "smooth" surface expression of the main fault. But significant variations of D appear along the main fault and show the presence of heterogeneities. D changes quite a lot between short and long wavelengths, the boundary being around 1 or 2 km. Short wavelengths have a higher value of D than longer wavelengths. The fault's surface may therefore be considered as close to a plane, and the roughness appears only at small scales. The conclusion of their study is that the southern segments, where D is higher, are rougher and thus less susceptible of a large earthquake, because they are too irregular to break during a single event.

The second article, by Okubo and Aki (1987) applies a box-counting technique to the same site. A minimum number N(r) of circles of radius r is chosen to cover entirely the surface expression of the San Andreas fault (Figure 2.17). The log-log plot of N(r) as a function of r allows the computation of the power-law exponent D. The global dimension for the San Andreas fault system is 1.31 ± 0.02. Dividing it into six segments, the authors show that the local values increase from the northeast, where D = 1.2, to the southeast, where D = 1.43. The authors link the variations of D in the segments to the variations of seismicity observed by Allen (1968), and make some points about the relationship between seismicity and the fractal geometry of faults. One important point is the detection of a break in the log-log graph for a critical radius r_c. This break implies that the mapped expression of the fault is not self-similar.

Okubo and Aki note that, if the faults' fractal nature persists so that critical lengths remain fixed in time, then it might be possible to anticipate some of the characteristics related to the degree of complexity of the fractal geometry of the fault trace. Although limiting themselves to the geometry of this surface trace, the authors suggest that the measures of complexity reflect the complexity of the fault surface itself, as:

$$D_{surface} = D_{trace} + 1 \qquad (2.30)$$

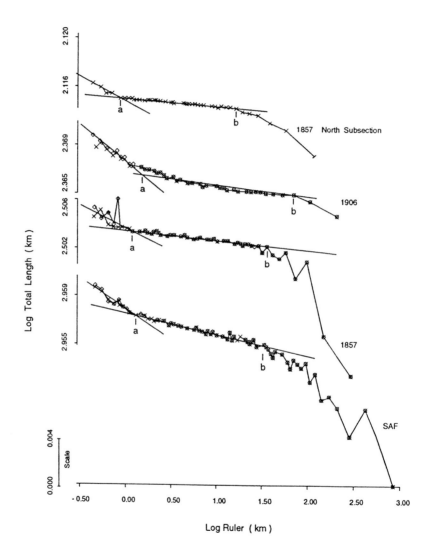

Figure 2.16. Method of Aviles et al. (1987). The authors apply Richardson's method to find the fractal dimension of the San Andreas fault.

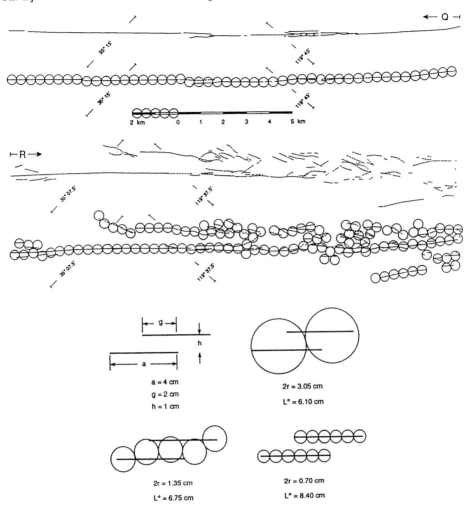

Figure 2.17. Method of Okubo and Aki (1987). It consists in covering with circles the faults associated to the San Andreas fault.

2.2.4 Percolation

The study of percolation shows the application of these techniques to physical properties of heterogeneous systems. The term "percolation" was introduced by Broadbent and Hammersley (1957) to describe fluid circulation in a porous medium. The following paragraphs will refer to recent works by Gilabert et al. (1990) and Sornette (1990).

The percolation concept can be explained by a simple example. Let us imagine two continental masses, very far from each other and separated by an ocean and a series of small islands isolated from each other. If the sea level declines progressively, the emerged portions of islands will grow and some will link together to form clusters. Beyond a limit sea level, enough islands will be connected to allow dry-land passage from one continent to the next. This transition between a series of unconnected islands, where the islands' maximum size is finite, to the merging of some islands into an infinite continent, is called percolation threshold (de Gennes, 1978).

Figure 2.18 shows this limit situation for the percolation threshold in a fissured medium. The model is similar to the model of the islands connecting one continent to the other. Here, the liquid can pass from one side to the other, through the fissured medium, when the percolation threshold is reached.

These properties can be formalised. S_i is the emerged surface, i.e. the islands' surface, and S_t the total surface of the islands and the ocean. The percentage of emerged lands is $p = S_i/S_t$. The value of p at the percolation threshold (when dry passage is possible) is noted p_c. When $p < p_c$, the length ξ is defined as the size typical of the largest cluster, i.e. the correlation length of the system. When $p > p_c$, there is always a continuous path linking the two continents. And when $p = p_c$, the percolation threshold is reached, and corresponds to the case where the size of the largest cluster becomes infinite. Otherwise said, the correlation length is infinite at the percolation threshold.

There are three important properties:

1. The percolation threshold always occurs for the same value of p_c, whatever the random drawing, for the same geometry and in the limit of the system.

2. This behaviour is similar to the critical phase transition, and is translated with a power law. If ξ is the correlation length, when $p \to p_c$, $\xi = \xi_0 (p - p_c)^{-\nu}$, where ξ_0 is of a size similar to the elementary object size, ξ diverges at ∞ when $p \to p_c$.

3. Close to p_c, ν only depends from the space dimension, and not from the system's geometry (contrary to p_c, which depends from the system's geometry). The independence of ν makes it a "universal" factor. In a two-dimensional system, $\nu = 4/3$ (Deutscher et al., 1993). For the preceding example, $\xi = \xi_0 (p - p_c)^{-4/3}$, where ξ_0 is the elementary initial size of an island.

This last point can be extended to other parameters than the correlation length, e.g. the conductivity or the breaking strength of two-dimensional fissured systems (Figure 2.19). In a heterogeneous system where the electric current is being studied, and where a cluster population can hamper the passage of current by its size, topology and structure, the electric conductivity G can be expressed as $G = G_0 (p - p_c)^t$. In a two-dimensional space, studies showed that $t = 1.3$ (Normand et al., 1988). In a fissure field, the breaking strength F of the system can similarly be expressed as $F = F_0 (p - p_c)^f$. And f is found to be: $f = 2 \nu = 2.66$ (Guyon et al., 1987).

These examples underline the unifying character of percolation as a critical system. Other properties of percolation will be examined in Annex B.2.

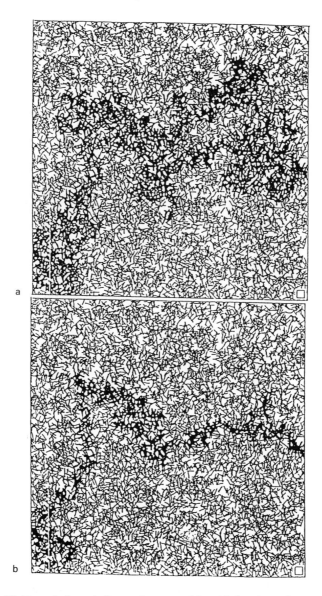

Figure 2.18. Percolation. A fissured system (size 13.5 cm), at the percolation threshold, is computer-generated. The size of the fissures is 3.5 cm, there are 8500 of them at the percolation threshold (i.e. when a continuous pathway exists between the two opposite sides of the system). In darker tones, we can see in (a) the infinite cluster which connects the two sides, and in (b) the skeleton of the infinite cluster obtained by removing the dead ends of the infinite cluster. From Vanneste et al., 1990.

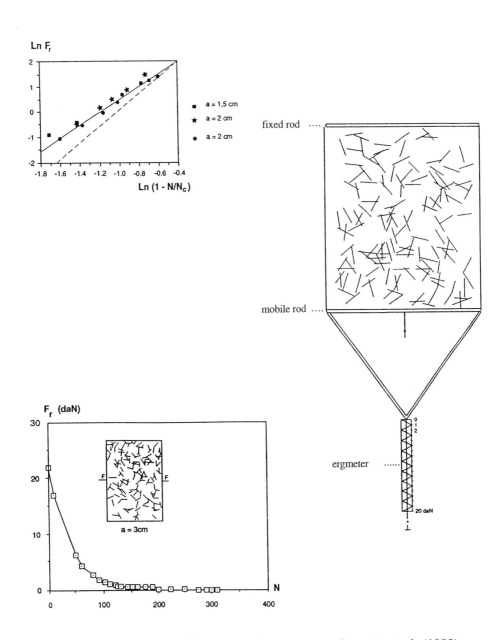

Figure 2.19. Analogy with the percolation process. Sornette et al. (1990) studied the relation: $F = F_0 (p - p_c)^f$. The value of the exponent f is found to be: $f = 2 \nu = 2.66$ (Guyon et al., 1987; Gilabert et al., 1990).

2.3 SEISMICITY - GUTENBERG-RICHTER LAW

It has been well known, at least for the last half-century, that there is a log-log relation between the number N of earthquakes in a given region and their energy (Ishimoto and Iida, 1939; Gutenberg and Richter, 1954). Starting with the definition of the magnitude of an earthquake, in 1935, Richter first formulated the law now known as the relation of Gutenberg-Richter:

$$\log.N = a - b\,m \qquad (2.31)$$

where N is the number of earthquakes of magnitude greater than m, and a and b are constants (see Figure 2.22).

Before linking this relation to the notion of scale invariance, as was done by Aki (1981), it will be interesting to develop further the notion of power-law distribution, which often leads to questions. In the examples presented here, such distributions are observed for the series of earthquakes of magnitude $\geq$ m (Gutenberg-Richter law), or for the time intervals between two events (eruptions, earthquakes, magnetic inversions, etc.) (see Sections 2.5 and 2.8).

2.3.1 Power Law or Poisson Law ?

A few tests allow us to determine the random or non-random nature of a time series. They are based on recognising the types of time distributions associated with the events.

<u>Histograms and cumulated distributions</u>

For all cases, we shall examine a theoretical series of random numbers, a theoretical triadic Cantor series, and, for comparison, a series of time data.

Normal distribution. The histogram of the normal distribution of the random variable x is bell-shaped, and is called a Gaussian. Figure 2.20 shows histograms built upon series of 100 and 1000 random numbers (drawn with an algorithm written by Khoklov, 1993). One can note the bell-shaped curve is better defined for the longest series. The cumulated distribution gives the three graphs of N, total number of values $\geq$ x, as a function of x; log.N as a function of x; and log.N as a function of log.x. Notice the linear aspect of the semilog curve, characteristic of a normal distribution of exponential or Poisson type. This is the test bringing into evidence the randomness of a numerical series.

Triadic set. The histogram computed on the series of "voids" in a triadic Cantor set after 7 iterations (series of 127 points) is very different from the previous one. It is shown that its envelope is exponential. The graphs of the cumulated distribution are very different from the previous ones. The log-log graph is a straight line, which is characteristic of a power-law distribution. The slope of the line is the exponent of the power law (Figure 2.21).

108   Applications to Geophysics   [Ch. 2

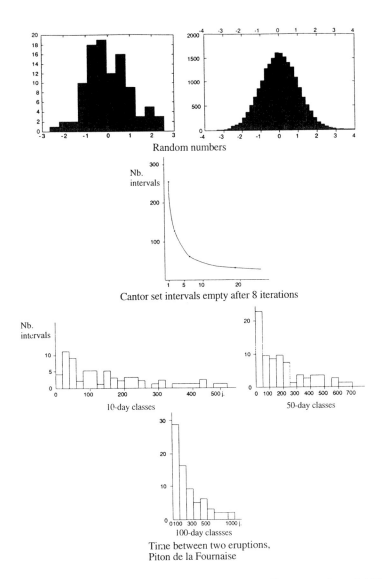

Figure 2.20. Examples of histograms, corresponding to series of 100 and 1000 random numbers, a triadic Cantor set, and the series of eruptions at the Piton de la Fournaise (Réunion) from 1930 to 1994 (in 100-day classes)

Observations. This example uses a series of volcanic eruptions on the Piton de la Fournaise (Réunion), with the time interval between the beginning of successive eruptions as a variable. The histogram (Figure 2.20) has a shape similar to the histogram of the Cantor set. The graphs of the cumulated distributions are also close to the linear log-log curve of the Cantor set. This similarity was observed during a preliminary analysis (Dubois and Cheminée, 1988) and had led us to propose the model of a power-law distribution rather than a Poisson-type model (Wickman, 1966, 1976; Klein, 1982; Kono, 1973). These authors deduce from their analyses (often conducted on too few data points) exponential distributions implying a random process. Contrary to them, we bring into evidence a power-law distribution which may imply determinism in the dynamic system which generates the eruptions at this volcano. We shall delve further into this point in Section 2.5 when looking at the probabilistic aspect of these methods.

The Gutenberg-Richter law expresses the same property, i.e. the power-law distribution of earthquake magnitudes. Indeed, magnitude is the logarithmic expression of energy, and the log-log graph shows a linear correspondence between the logarithm of the cumulated number of earthquakes of magnitude $\geq$ m and this magnitude.

This was linked to the notion of scale invariance and fractal object by Aki (1981).

The relation was improved by introducing the seismic moment. Hanks and Kanamori (1979) established that:

$$\log.M_0 = d + c\, m \tag{2.32}$$

where M_0 is the seismic moment, m is the magnitude. d and c are constants, with values of c = 1.5 and d = 16 (Thatcher and Hanks, 1973; Purcaru and Berckhemer, 1978; Hanks and Kanamori, 1979).

Kanamori and Anderson (1975) had found the proportionality relation between the seismic moment and the fault length (or dimension):

$$M_0 = k\, L^3 \tag{2.33}$$

Combining the preceding relations, we get:

$$\log.N = -\frac{3b}{c} \log.L + \log.f \tag{2.34}$$

or:

$$N(L) = f\, L^{-3b/c} \tag{2.35}$$

with:

$$f = \frac{bd}{c} + a - \frac{b}{c} \log.k \tag{2.36}$$

When this relation is verified, the slope of the graph of {N(L); log.L} is -3b/c, and b and c respectively are the slopes of the semilog graphs {frequency; magnitude} and {seismic moment; magnitude}.

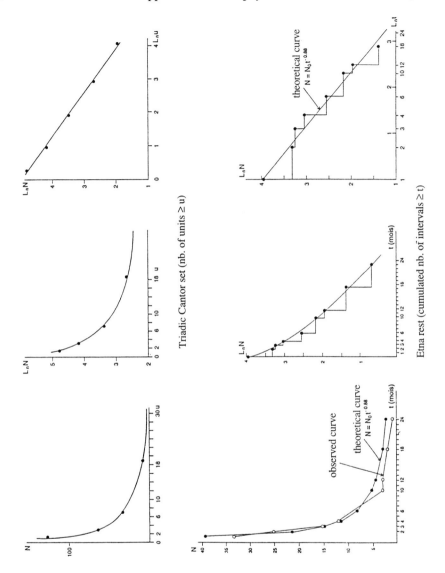

Figure 2.21. Cumulated distributions. Normal, semilog and log-log graphs of N as a function of x are shown for the three cases examined in the previous figure: random numbers, triadic Cantor set and series of eruptions at the Piton de la Fournaise.

If we compare with the definition given by Mandelbrot (1975, 1982) of the number of fractal objects with a size greater than r:

$$N \sim \alpha r^{-D} \tag{2.37}$$

we can deduce that the fractal dimension of a seismic fault can be written as D = 3b/c.

Aki (1981) considers three groups of possible mechanisms for a series of earthquakes in a given region, according to the value of b.

- If b = 1, the fault dimension is 2 (as c ≈ 3/2), i.e its topological dimension.
- If b = 3/2, the dimension is 3. This corresponds to a series of earthquakes whose fault planes tend to fill a volume.
- It is also possible to have values of 0.5 < b < 1, corresponding to fractal dimensions between 1 and 2. In this case, the fault cannot be considered as a plane, but as rupture lines tending to fill the plane. This situation seems to correspond to the Goishi model of Otsuka (1972) and Maruyama (1978). The branching model of Vere Jones (1976), consisting in the propagation of a rupture, has a geometry corresponding to micro-fissures in the medium and the propagation of the rupture along the branches of the micro-fissures.

In this approach, the study of objects such as earthquakes is using a limited number of parameters amongst the 5 available (as a first approximation): 3 spatial coordinates, time, and the energy produced (the earthquake's magnitude). Here, we only use earthquakes occurring in a given region and exceeding a given energy. The time parameter is not analysed as extensively as it will be in later studies of the probabilistic aspect, on its own for the Cantor dust, or with other parameters in the Russian methods (see Section 2.7).

The computation of D for several types of earthquakes allows us to better understand the possible mechanisms. The study of the variability of b has given rise to several works (e.g. Mogi, 1962; Scholz, 1968), on laboratory samples as well as in the field. Main (1987) worked on seismicity models before the Mt. St. Helens eruption of 1980. He uses the similarity dimension with the method we have described in the previous paragraphs. Immediately before and after a large earthquake, the value of b changes from b < 1 to b > 1. This is explained by the releasing of stress on the main branch of the self-similar fault system. Before the break, b < 1 and D < 2. After the release around the main fault, only secondary branches participate to the release of residual stress, yielding higher values of b > 1 and D > 2.

It has also been remarked that a decrease in b may be the precursor of large earthquakes (Smith, 1981), or of the breaking of a rock sample in laboratory experiments (Ohnaka and Mogi, 1982), or of a volcanic eruption (Gresta and Patane, 1987; Yuan et al., 1984).

This technique was applied by Nouaili et al. (1987) and Dubois and Nouaili (1989) to subduction zones by dividing the seismic zones into three regions: 100 - 300 km depths, 300 - 500 km, and 500 - 700 km. The different subduction zones will exhibit different behaviours and so will the different depths inside the same zone. These results are based on the investigation of 11 subduction zones around the Pacific Ocean (Figure 2.22). It starts from the realisation that the branching model, characterised by 1 < D < 2, is the most frequent process in subducting lithospheres.

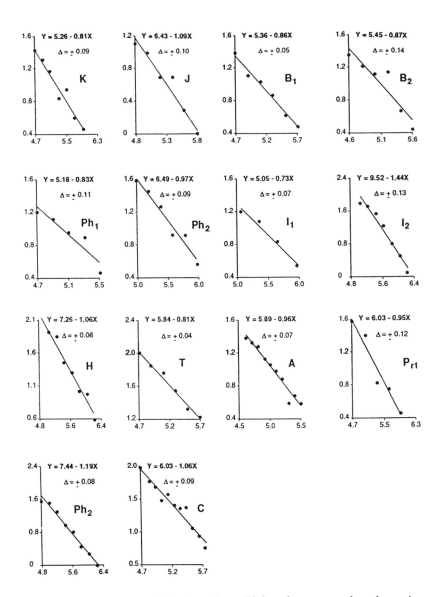

Figure 2.22. Illustration of the Gutenberg-Richter law, on earthquake series in the different subduction zones around the Pacific Ocean, with focal depths between 100 and 300 km. K, J, B, Ph, I, H, T, A, Pr, and C respectively design the zones of Kermadec, Japan, Bonin, Philippines, Java, New Hebrides, Tonga, Aleutians, Peru, and Chile. Δ is the standard deviation on b, $Y = \log.N$, $X = m$ (Dubois and Nouaili, 1989).

This confirms the mechanisms proposed by some seismologists (e.g. Frohlich and Willemann, 1987), who remark that earthquakes do not occur preferentially along the nodal planes of the main shocks, but that the breaks occur in a non-planar zone distributed around the initial focal point. The value of D implies a stress regime associated with the dipping of dense lithosphere into a viscous and less dense asthenosphere. This is for example the case for the Indonesia-Banda Sea subduction zones, with lateral stress adding to the main contribution from subduction. High values of D appear at the edges of the dipping lithosphere, in the Tonga Islands. They express the fact that the rupture planes of all earthquakes tend to occupy the whole space. This supplements the observations of Giardini and Woodhouse (1984) of a north-south deformation of the edge of the lithosphere, where the mesosphere introduces lateral stresses which complicate focal mechanisms between 500 and 700-km depths. Conversely, in the contact zone between the subducting and subducted plates, the values of D are generally close to 2, implying a simpler geometry of the contact area (Tonga Islands, cf. Nouaili et al., 1987). In summary, the variability of D along the subducting lithosphere is a good expression of the fracturation processes at play in the lithosphere, and our interpretations of it are in agreement with the geodynamic setting of the principal circum-Pacific subduction zones.

2.4 FRAGMENTATION - TECTONICS - SEISMICITY

The studies of these different processes linked to rock fragmentation and fracturation have been linked in a recent study by Nagahama and Yoshii (1994). Long before the beginning of the fractal approach, power laws had been established relating the different physical parameters (Gaudin, 1926; Schuhmann, 1960; Bond, 1952; Bergstrom et al., 1963; Charles, 1957; Tartaron, 1963; etc.). Nagahama and Yoshii (1994) investigate the different scaling laws intervening in fracturation.

The first one expresses the power-law dependency of the cumulated number N(r) of fragments with a size larger than r:

$$N \sim \alpha r^{-D_S} \tag{2.38}$$

where D_S is the fractal dimension of the fragment size distribution (see Section 2.2).

The second scaling law quantifies the amount of roughness of breaking surfaces in $\mathcal{R}^3$. This dimension D_R is the fractal dimension of the surface's shape (see section 2.1). Its value ranges from 2, when the surface is smooth, to 3, when the surface is so complex that it tends to fill a whole volume (Mandelbrot, 1977, 1982).

It was also shown that the cumulated mass M(r) of fragments with a size smaller than r can be expressed as a power-law:

$$M(r) \sim r^h \tag{2.39}$$

where h is a positive constant (Gaudin, 1926; Schuhmann, 1940; Fujiwara et al., 1977; Turcotte, 1986).

If all fragments from a given set have a common fractal dimension D_R for their roughness, their cumulated mass becomes (Nagahama and Yoshii, 1994):

$$M(r) = \int_0^r r^{D_R - D_S - 1} \, dr \sim r^{D_R - D_S} \qquad (2.40)$$

Comparing equations (2.39) and (2.40), we get:

$$h = D_R - D_S \qquad (2.41)$$

Nagahama and Yoshii then use the energy per mass unit necessary for fragmentation. According to Avnir et al. (1983):

$$E \sim r^\omega, \ \omega = D_R - 3 \qquad (2.42)$$

From the empiric law of Walker-Lewis, we have:

$$E = C \, r^{-n+1} \qquad (2.43)$$

where C and n are constants (Bergstrom et al., 1963).

Charles (1957) established that $h - n + 1 = 0$, and therefore, from equations (2.42) and (2.43), we get: $\omega = D_R - 3 = -n + 1$. And, as $h = D_R - D_S$,

$$\omega = h + D_s - 3 = D_S + n - 1 - 3 \qquad (2.44)$$

The preceding equations give: $2\omega = D_S - 3$ and $D_R = \frac{1}{2}(D_S - 3)$, i.e.:

$$D_R = \frac{1}{2}(D_S + 3) \qquad (2.45)$$

Equation (2.45) means that the fractal roughness dimension D_R in a 3-D space equals the mean value of the fractal dimension of the size distribution, D_S, and the Euclidean dimension of the space in which the observation is conducted (here, this dimension is 3). This shows there is a constraint in the fractal geometry of fragmentation.

If we consider the lithosphere is a fragmented object, and that tectonic faults correspond to the limits of these fragments, the roughness of fault surfaces can be easily computed.

Using the law of Gutenberg-Richter, other relations have been proposed by Nagahama and Yoshii (1994).

It was demonstrated by Aki (1981) that (see Section 2.3):

$$D_S = \frac{3b}{\delta} \qquad (2.46)$$

where b is a constant from the Gutenberg-Richter law and δ a constant related to the relative durations of the seismic source and the time constant of the recording system.

Figure 2.23. Earthquakes and volcanic eruptions seem to be distributed randomly in the time domain, as here during a seismic event in the White Cordillera, Peru (J. Deverchère, personal communication). Numerical analysis proves in fact that the distribution is far from random (see text for details).

In most seismologic studies, a value of $\delta = 1.5$ is selected (Kanamori and Anderson, 1975). This yields: $D_S = 2b$ (see Section 2.3). This last equation shows that the values of b are related to the fractal dimension of lithospheric fragmentation.

Going back to the roughness dimension, the equations (2.46) and (2.45) can be combined into:

$$D_R = \frac{3}{2}\left(\frac{b}{\delta} + 1\right) \qquad (2.47)$$

This applies to the roughness of a seismic break zone. The authors deduce that this relation will lead to important developments in our knowledge of the breaking process, and hence of earthquakes, inside a tectonic field (Nagahama and Yoshii, 1994).

2.5 PROBABILISTIC APPROACH - APPLICATIONS

2.5.1 Observables

The application of fractal analysis to probabilist ends was one of the reasons for the introduction of fractals by Mandelbrot (1975, 1982, 1989). We shall see its extension to seismicity and series of volcanic eruptions, using the Cantor dust methods, the correlation function and the first-return map.

Let us start from a truncated application of the Cantor dust method, with internal and external scales verifying $e > 0$ and $E < \infty$. The order of the intermissions (empty intervals between two events) is randomised, i.e. drawn randomly to make them statistically independent. If u is the width of an intermission, the length distribution satisfies:

$$P_r(U \geq u) = u^{-D} \qquad (2.48)$$

In other words, the probability of reaching or exceeding u is u^{-D}. In a time series where events (earthquakes or volcanic eruptions) have been identified, and where a power-law of exponent D has been found (on a log-log graph), the probability that a new event i+1 happens at a time t after the event i is t^{-D}, $0 < D < 1$.

The clustering of events will be shown to depend on D; the smaller D, the larger the tendency of events to cluster together.

The clustering threshold is chosen as t_0. The intermissions "$> t_0$" and "$< t_0$" are separated and their relative durations are divided by t_0. When D is small, the relative durations of intermissions $> t_0$ have a high probability of being larger than their lower limit 1. Indeed, the conditional probability that $T > 5 \, t_0$ is 5^{-D}, and therefore tends toward 1 when D tends toward 0. But the relative durations of intermissions $< t_0$ become smaller than 1. Events therefore tend to cluster around t_0, the more so when D is small. Conversely, when D is close to 1, intermissions tend to be more regular.

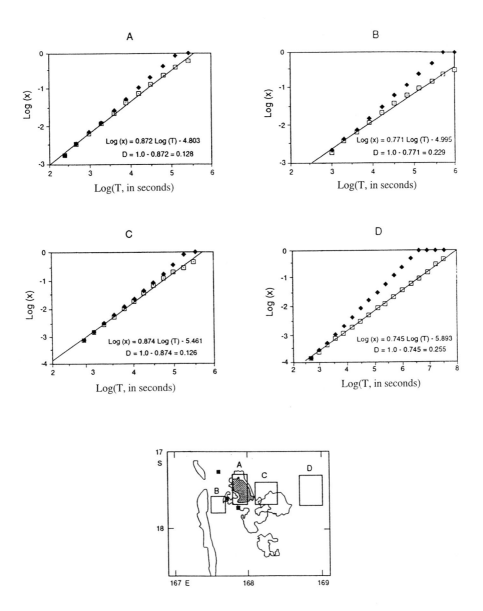

Figure 2.24. Fractal dimension of seismicity in the time domain. Computed on earthquake series in Vanuatu, the value of D varies with the study area (Smalley et al., 1987).

This probabilistic approach was applied by Smalley et al. (1987) in the course of a study of the evolution with time of seismicity in several regions of the New Hebrides. They first showed that the distribution law of intermissions was significantly different from a random Poisson-type distribution, and that it was fitting a power law, i.e. a fractal law. The observed values of D range from 0.126 to 0.255, according to the study area. They show that some regions are more marked by clusters of earthquakes than others (Figure 2.24).

Smalley et al. (1987) underline the very preliminary character of this approach, as seismic events possess 5 dimensions (time, the 3 spatial dimensions, and magnitude), and as clustering can in principle be analysed in any subset of these. The choice of time as a parameter was only made to look empirically at the possible correlation between the variations of D and the triggering of a major earthquake.

2.5.2 Application to Volcanic Eruptions

When observing the time series of eruptions for a particular volcano, the different events seem strikingly irregular and impossible to quantify. Several authors (Wickman, 1976; Klein, 1982) have noted that the eruptive behaviour of most volcanoes was indistinguishable from a random behaviour, which apparently precludes any possibility of developing efficient predictive methods.

But the random character of eruptive series has never been systematically analysed with appropriate tests. For Wickman (1976), the time intervals between successive eruptions seem distributed along simple statistical laws (exponential distributions) or, for other volcanoes like Vesuvius, by successions of different volcanic stages (Markovian chains). As for Klein (1982), he uses Kolmogorov-Smirnov tests.

A recent study (Sornette et al., 1992) showed that bringing into evidence the random nature of a particular process in time, requires several distinct tests. These tests can be divided into two main categories.

The first category of tests shows that the events are following a well defined probability distribution, such as Kolmogorov-Smirnov. The distribution may be flat, corresponding to a uniform probability, or have any other shape.

The second category of tests mainly detects correlations in the series of events. For example, the sequential correlation test computes a usual correlation function which, when it disappears, indicates that the events are independent of each other. Another test consists in examining the lengths of monotonous series and comparing the distribution of segment lengths to the ones that would be observed for a purely random variable. A test, called the "sub-series" test, applies the last test to a sub-series obtained, for example, by extracting one event from every couple of events in the initial series.

However, a very important fact should be noted. Arnéodo and Sornette (1984) recognised recently that some determinist systems verify all tests on the random character of the events they generate. This means that a positive answer to all tests on the randomness of a series does not imply that the system is not determinist. These problems were mentioned in Chapter 1.

Two types of systems are identified:

1. The first type comprises all systems with several degrees of freedom which are intimately coupled and can develop a random dynamics, resulting from the complexity of superposition and coupling of the different evolutions of each degree of freedom (e.g. Brownian movement). The evolution of the physical variable studied is indistinguishable from a random process. Any prediction is completely illusory. In Section 1.7.4, this was called the chaotic chaos.

2. The second type comprises systems with a small number of degrees of freedom, which may present very complex dynamic behaviours, and have been defined as related to determinist chaos.

The main conclusion of these discussions is therefore that the observation by Wickman (1976) and Klein (1982) that the time series of volcanic eruptions verify the random tests outlined, does not obligatorily imply that their generating systems are not determinist. This is why we have applied three new tests to these time series; the Cantor dust method, the Grassberger and Procaccia's correlation function and the first-return map. These methods were developed in Chapter 1, and we shall therefore focus on their results.

2.5.3 Cantor Dust

This method has been applied to datasets from several volcanoes (Dubois and Cheminée, 1988, 1991). In the case of basaltic-type volcanoes such as the Piton de la Fournaise or Hawaii, the fractal dimensions show two characteristic values: $D = 0.45$ for rest periods between 1 and 10 months, and $D = 0.67$ for rest periods between 12 and 48 months (Piton de la Fournaise). For Kilauea, the respective fractal dimensions are $D = 0.58$ for rest periods below 2 years and $D = 0.75$ after 2 years. For Mauna Loa, $D = 0.44$ for periods below 6 years and $D = 0.84$ for periods longer than 6 years (Figure 2.25).

This double periodicity was interpreted as an expression of the filling and emptying of one (or several) magmatic chambers with distinct flow rates.

For both cases, Piton de la Fournaise and Hawaii, the smallest value of D corresponds to a tendency of eruptions to cluster. This seems related to the superficial mechanic response of the edifice to higher pressures created by magma cooling in a shallow reservoir (Tait et al., 1989). The second, larger, value of D corresponds to a more regular rythm and, for the Piton de la Fournaise, may be related to picrite eruptions at depth.

Conversely, a basaltic volcano like Etna shows only one regime, and a study for the period 1930-1987 determined a single value of $D = 0.85$, corresponding to a regular functioning.

When applied to subduction zone volcanism, this method shows a large variability of the fractal dimension D. But, generally, andesitic volcanoes of Peleian type (Mt. Pelée, Soufrière de la Guadeloupe, Mt. St. Helens, Fuji) have low values of D, $0.2 < D < 0.4$. Indonesian volcanoes, with their frequent and important activity, have values close to 0.7 (except for Krakatau, for which there is insufficient data).

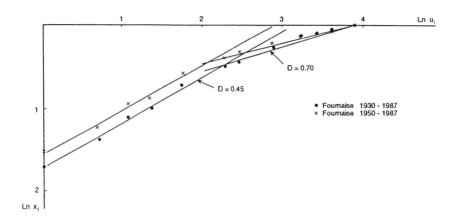

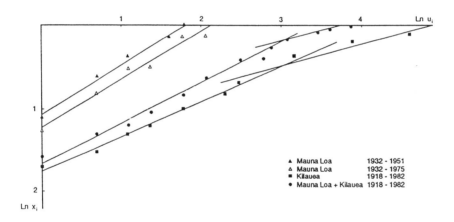

Figure 2.25. Fractal analysis using the Cantor dust technique. The data shown here come from time series of volcanic eruptions at the Piton de la Fournaise (Réunion) and Hawaii. From Dubois and Cheminée, 1991.

2.5.4 Correlation Function

The main difficulty for applying this technique is the number of points necessary. The use of the correlation function with short series was detailed in Chapter 1, and it was demonstrated that it was possible to use series of down to 100 points when the dimension of the attractor was less than 3.

For this reason, the technique could only be applied to three volcanoes: Piton de la Fournaise, Hawaii and Vesuvius (Dubois and Cheminée, 1991; Sornette et al., 1991). The analysis uses the three variables $t_1(i)$, $t_2(i)$ and $t_3(i)$. $t_1(i)$ is defined as the time interval between the end of eruption i and the beginning of eruption (i+1). $t_2(i)$ is the duration of the eruption i. $t_3(i)$ is the time interval between the beginning of eruption i and the beginning of eruption (i+1). Of course:

$$t_3(i) = t_1(i) + t_2(i) \tag{2.49}$$

Figure 2.26 sums up the results for the Piton de la Fournaise volcano. The $t_2(i)$ series appears distinctly random (increasing with d, dimension of the phase space). The saturation for $t_3(i)$ in the phase space around the value of 1.7 confirms the results obtained with the Cantor dust method. This test shows that the $t_1(i)$ series results from determinist dynamic system with an attractor of dimension 1.7. The dynamic system therefore has 2 degrees of freedom, which confirms the double periodicity observed previously.

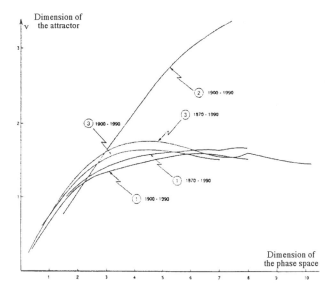

Figure 2.26. Application of the correlation function method. The results shown here concern the series $t_1(i)$, $t_2(i)$ and $t_3(i)$. $t_2(i)$ appears to be random (v increases continuously with the dimension d), $t_1(i)$ is noisy but, for $t_3(i)$, saturation appears for $v = 1.7$. From Dubois and Cheminée (1991).

The application of the correlation function test to the two active Hawaiian volcanoes is less easy. Each volcano, Kilauea and Mauna Loa, has only had a limited number of eruptions during recent history (38 for the former, 46 for the latter). To apply the test, we merged the two series into one dataset susceptible of giving us details about the dynamic behaviour of the whole Hawaiian complex (Sornette et al., 1991). The resulting value was $D_{max} \approx 4.6$, which seems to correspond to a dynamic system with more than twice the number of degrees of freedom for the Piton de la Fournaise. This may result from two independent dynamic systems, one controlling the Kilauea eruptions and the other the Mauna Loa eruptions, and their respective dimensions being close to 2. These two possible systems should be weakly coupled to avoid the coupling of phases, which would reduce the attractor's total dimension from the sum of the 2 dimensions to a smaller value indicative of the degree of coupling.

A third test of this method was performed on a relatively long series of 160 eruptions of the Vesuvius (Dubois and Cheminée, 1991). For the interval 1694-1872, an acceptable saturation is obtained for $D_{max} = 2.9$. This implies that the dynamic system generating the eruptions has 3 degrees of freedom, which is consistent with the results from other approaches (e.g. Carta et al., 1981).

2.5.5 First-Return Map

This method was described in Chapter 2, and consists in plotting the couples of points (x_{i+1}, x_i) of the series:

$$x_i \Big|_{i=1}^{\mu} \qquad (2.50)$$

In the present case, we use $t_3(i)$, the series of time intervals between the beginning of eruption i and the beginning of eruption (i+1), for the Piton de la Fournaise (Sornette et al., 1991). Figure 2.27 shows the points t_{i+1} plotted as a function of points t_i.

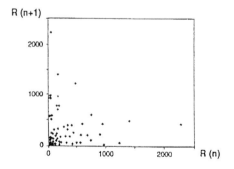

Figure 2.27. Example of a first-return map. This test was used on the t_3 series computed for the Piton de la Fournaise. The result shows a situation intermediate between a purely random distribution and a determinist distribution (Sornette et al., 1991).

If the points with coordinates (t_{n+1}, t_n) were distributed along a simple curve, this would mean that t_{n+1} is a simple non-linear function of t_n, i.e. that the dynamics behind the series of t_n is a simple 1-D determinist law. On the other hand, if the series was completely random, the (t_{n+1}, t_n) plane would be filled densely and uniformly. Figure 2.27 shows that the current situation is intermediate.

The dimension of the attractor for this series (1.7) was found earlier with the correlation function. It would therefore be illusory to try to represent simply this curve. One way to proceed would be to find a suitable Poincaré section. For an attractor constructed on a series of events, the equivalent of a Poincaré section consists in extracting sub-series of the whole series (cf. Bergé et al., 1984, about the Rössler attractor).

In this case, Sornette et al. (1991) observed on Figure 2.27 that most points close to the origin seemed due to noise. This suggests that two eruptions occurring too close together cannot be distinguished and that they may correspond to the same higher stage of the volcano's dynamic evolution. In these conditions, sub-series can be extracted after eliminating the eruptions verifying $t_i - t_{i-1} < t_{min}$. Several trials, for values of t_{min} between 100 and 200 days, allowed the "cleaning" of the data to produce clearer graphs (Figure 2.28).

A second way of processing the data consists in selecting the eruptions i for which t_{i-1} and t_{i+1} are smaller than t_i. The graph of the remaining values (24 out of 72) is shown in Figure 2.28. The presence of a curve that fits well the resulting points is characteristic of a determinist chaos (Bergé et al., 1984).

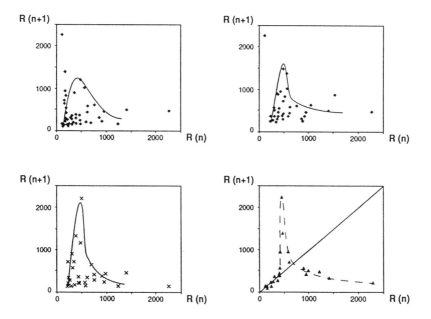

Figure 2.28. First-return maps obtained with sub-series. In these sub-series, the eruptions too close to the previous one were removed from the dataset, with periods: (a) $R_{min} = 100$ days; (b) $R_{min} = 200$ days; (c) $R_{min} = 200$ days and the events (i-1) and i have been removed as well; (d) only the longest series was kept (Sornette et al., 1991).

A similar approach was used on data from the Hawaiian volcanic complex, whose attractor's dimension was larger. The resulting curve is different from the one for the Piton de la Fournaise and shows two maxima (Figure 2.29). This may be due to the high value of its attractor ($D \approx 4$), suggesting the interaction of two different non-linear dynamic determinist systems.

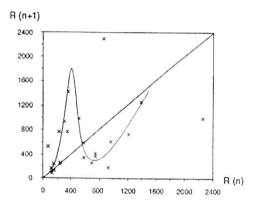

Figure 2.29. First-return map for the eruptive series of Hawaii. The same procedure as in Figure 2.28 was used. The presence of two maxima may be explained by the higher value of the attractor, $D \approx 4$.

Last, this method was applied to recent series from the Anak Krakatau, a young volcano emplaced in 1929 on the site of the old volcano which exploded in 1883 (Dubois et al., 1992). The simplicity of the resulting curve shows that the attractor's dimension is certainly close to 1, and hence that the current eruptive mechanism is linked to the evolution of a single, shallow, magma chamber, relict from the large 1883 eruption. This confirms the hypotheses generally made by volcanologists (e.g. Camus and Vincent, 1982, 1987).

It is also worth noting an important study performed by Shaw (1988) on Kilauea, using the first-return map on the volume of Kilauea's summit reservoir.

2.5.6 Multifractal Analyses

Shaw and Chouet (1989) applied multifractal analysis to a series of tremors recorded at Kilauea over a period of 22 years. The method used is the FSA (Fractal Singularity Analysis) of Meakin et al. (1985, 1986). It starts with a first-return map where the duration x_{n+1} of an incoming tremor is plotted as a function of the duration x_n of the present tremor. This method is detailed in Chapter 3, and consists in estimating the probabilities from the recurrence times $m_i = p_i^{-1}$; m_i being the number of steps needed to come back to a distance l from a given start point.

The values of m_i are determined for one value of l at each point, elevated to the power (1-q) and averaged. This is done for all points, for several values of l and q, and allows the computation of the slopes τ, for q constant, of log-log graphs of the averages (m^{1-q}) as a function of l:

$$\alpha = d\tau/dq \qquad (2.51)$$

and:

$$f = q\alpha - \tau \qquad (2.52)$$

One then plots $f(\alpha)$ as a function of α (Figure 2.30).

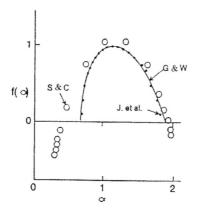

Figure 2.30. Application of FSA (Fractal Singularity Analysis) to a series of 577 tremors recorded in Hawaii for 22 years (Shaw and Chouet, 1987). The results are compared to the conclusions from several models: convection (Jensen et al. 1985), CGMN model (Golden Mean Number; Halsey et al., 1986).

For their study of Hawaiian tremors, Shaw and Chouet (1987) used 577 events recorded during 22 years. The uncertainties in the application of FSA show that α is unstable below 0.6 and above 1.5. This is due to the number of data points, small for this kind of analysis. Other studies benefited from more points (Figure 2.30): 2,500 points for the convection movements analysed by Jensen et al. (1985), 70,000 points for the electronic transport processes analysed by Gwin and Westervelt (1987).

It is worth noting that multifractal analysis can be applied to many domains. Seismicity is one example, with the construction of singularity spectra in the time domain, at a given location (interval between earthquakes), or in the spatial domain, at a given time (Geilikman et al., 1990).

2.6 SELF-ORGANISED CRITICALITY: SOC

The concept of self-organised criticality (SOC) was introduced by Bak, Tang and Wiesenfeld in 1987 to interpret the 1/f noise. Several articles then showed the interest of SOC for the modelling of seismic events and the structure of the lithosphere (Bak et al., 1987, 1988, 1989; Bak and Tang, 1989; Sornette and Sornette, 1989; Sornette et al., 1990; Bak and Chen, 1991).

Large natural interactive systems evolve toward a critical state in which a minor event can have dramatic consequences. Thus, the application of SOC to seismicity means that earthquakes contribute to organise the crust or the lithosphere in space and in time. This can explain the dynamics of earthquakes and of many other systems (natural or even economic).

The classic example was presented by Bak and Chen (1991) and represents a set of dominoes disposed along three different patterns (Figure 2.31): critical, sub-critical, and over-critical. In the critical state, the dominoes are placed randomly over approximately half the segments of a diamond-like grid (45°- diagonals). When the dominoes of the lower rows are pushed, the critical system undergoes chain reactions of various kinds. In the sub-critical system, the density of dominoes is smaller, and the chain reactions are smaller in scale. And, in the over-critical system, the density of dominoes is higher than in the critical state; the system "explodes" (nearly all dominoes fall down).

Another well-known example is proposed by Bak and Chen (1991). It consists in a sand heap, to which sand grains are continuously added, creating avalanches on its slopes. The sand heap is placed on a precision scale (precise to $\pm\, 0.0001$ g), with a maximum weight of 100 g (Held et al., 1990). Each grain weighs approximately 0.0006 g. A sand heap with a diameter of 4 cm at its base weighs approximately 15 g. It is therefore possible to follow the distribution of avalanche sizes in time, knowing the sand grains are added to the system with a regular rhythm (Figure 2.32).

At the onset of the experiment, the grains stay close to where they fall. After some time, they form a heap whose slope is shallow. The slope becomes steeper and, in some places, becomes so steep that grains slide and create small avalanches. The more sand grains are added, the steeper the slope, and the more important the avalanches. A few grains then start falling outside of the scale, and the heap stops growing when the amount of sand added to the system amounts to the amount of sand that falls out. This is the critical state.

If the experiment is continued past this critical stage, a new sand grain can provoke avalanches of any size, including catastrophic avalanches. But, in most cases, the sand grains fall in such a way that no avalanche occurs. Even the largest avalanches do not much modify the slope of the critical state. Let us examine more closely the triggering of an avalanche. A single sand grain will be stopped in its fall only if it falls onto a stable position, otherwise it will go on falling. If it collides with unstable grains, it will trigger their own fall and the process will continue, each moving grain stopping or falling further and colliding with other grains. The process stops when all active particles are stopped or have fallen outside.

Ch. 2] Self-Organised Criticality (SOC) 127

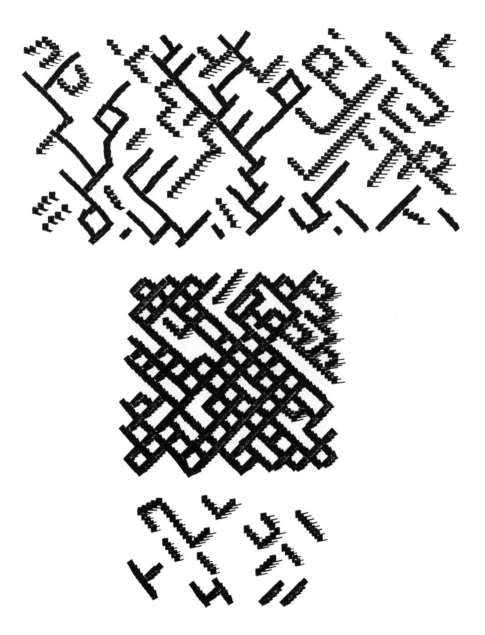

Figure 2.31. Self-organised criticality (SOC) and the game of dominoes. Three cases are possible: (a) critical; (b) sub-critical; (c) over-critical. From Bak and Chen, 1991.

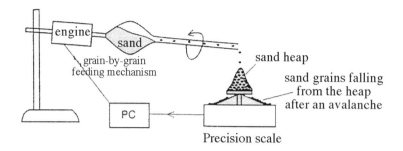

Figure 2.32. The classic example of SOC on a sand heap (Held et al., 1990). The sand grains fall one by one on top of the heap, with a regular rhythm. Avalanches occur and are studied with the precision scale.

The state of the system can be studied at each instant with the recording by the precision scale of the infinitesimal weight variations. The results show that mass fluctuations are scale-invariant (Figure 2.33), and that the probability distribution of avalanches looks like a finite-scale law. On Figure 2.33, the weight variations are represented as a function of the number of grains, and the close-up views from (a) to (b) and from (b) to (c) perfectly illustrate the scale invariance.

The application of the concept of self-organised criticality to the lithosphere and to earthquakes (Sornette et al., 1990) is based on the same idea, namely that the local slope cannot increase beyond a given threshold without triggering avalanches, which bring the local slope below the instability threshold. Sornette et al. (1990) proposed to examine a function $\sigma(S,\varepsilon)$ of the stress S and the deformation tensor ε. When $\sigma(S,\varepsilon)$ reaches a given threshold K, corresponding to the crust-breaking criterion (i.e. an earthquake), faults are created and release some of the stress. Thus $\sigma(S,\varepsilon)$ cannot grow above K. But at the same time, it cannot be much smaller than K, otherwise earthquakes would not occur constantly in large regions.

By analogy to the sand heap model, Bak and Tang (1989) suggest that the avalanches caused by the addition of new sand grains represent the earthquakes. When the pressure increases, the avalanches become bigger and bigger. At the critical state, there are no characteristic time, space and energy scales, and all the spatial and time correlation functions are power laws. Therefore, for Bak and Tang (1989), the Gutenberg-Richter law can be interpreted as an expression of the self-organised critical behaviour of terrestrial dynamics. The fractal geometric distribution and the dynamics of earthquakes are spatial and time signatures of the same process.

Sornette et al. (1990) take this analysis further. For them, the correlation functions of the deformation tensor define a number of critical exponents. The dynamic exponent Z rules over the spatial diffusion with time of the deformation variations. The field exponent describes the renormalisation of deformations when changing scale. And the spatial anisotropy exponents control the anisotropy of the diffusion of deformations in space. Solving the non-linear equation of generalised tensor diffusion would require the determination of these exponents. The authors acknowledge that they could not solve the whole problem, whereas Hwa and Kardar (1989) solved it for the sand heap example. Following the concepts they introduced, and the large-scale fractal organisation of faults inside continental plates, Sornette et al. (1990) cast some doubts about models of the location of seismic focal mechanisms on a single fault with branches only coupled at their extremity. Their field theory and their fractal analyses imply that stress and deformation are coupled on large distances.

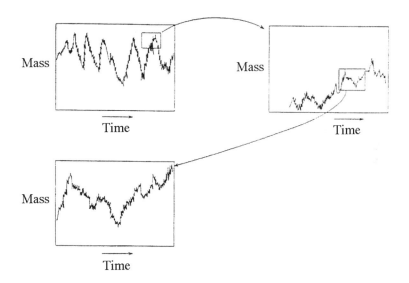

Figure 2.33. Results from the sand heap experiment. Scale invariance is illustrated by the evolution of the process in (a), (b) and (c), identical images at smaller and smaller scales.

2.7 THE RUSSIAN METHOD: ALGORITHMS M8 AND CN

This method is based on the statistical approach undertaken by the "Russian school" initiated by the research of Kolmogorov, and the concept of pattern recognition. The strategy developed by this group from Moscow's ITEP (International Institute of Earthquake Prediction Theory and Mathematical Geophysics) relies on two types of approaches for earthquake prediction: the phenomenological modelling and the theoretical modelling of earthquake flow, i.e. the time series of earthquakes in the space-magnitude domain. Each of these approaches is based on the concept of a lithosphere, source of most earthquakes, as an unstable non-linear system made of a hierarchy of interacting blocks and whose dynamics is a characteristic determinist chaos (Keilis-Borok, 1990).

Without going further into the description of these statistical analyses, we shall present the principles of the method, and two algorithms enabling the determination, in a known seismic area where enough data has been gathered, of the TIP's (times of increased probability) of earthquake occurrence.

2.7.1 Principles

The main ideas of this method were introduced by Keilis-Borok (personal communication, 1993). The response of a fault system to excitation increases before a large earthquake. It is shown in the earthquake flow by, for example, the increase of its intensity, irregularity in space and time, the clustering of earthquakes, and their domain of interaction. The method is based on the processing, still quite unrefined, of these events by the algorithms described below. The theory is based on four paradigms that result from global studies and numerical simulations of seismicity:

1. *Large-scale interactions*. If they exist, the signals of the occurrence of a large earthquake may not come from the vicinity of a future epicentre, but from a vast system of faults with different tectonic types (strike-slip, thrust faults, etc.).
2. The apparition of *vague precursor events*, each of which, if isolated, is not sufficient for prediction, but which, taken as a whole, are a more promising alternative than a single well-defined precursor.
3. The *similarity*, on a global scale, of precursor phenomena.
4. The *relevance* of non-linear dynamics, with its concepts of chaos and self-organisation.

A statistical study of predictions shows how relevant the concept of chaos is to seismic prediction. But it brings up a number of problems, in the control of parameters; in the search for the attractor's dimension; in the limits, similarity and self-similarity; and in the limits of the predictability.

2.7.2 The M8 Algorithm

The M8 algorithm is described in several articles (Keilis-Borok and Kossobokov, 1984, 1986, 1988, 1990; Gabrielov et al., 1986).

The earthquake flow associated with earthquakes of magnitude greater than or equal to a given threshold M_0 is defined first, even if these events are after-shocks of a larger earthquake.

The dimension of the area considered depends on the magnitude M_0 of the earthquake for which we try to determine the TIPs (times of increased probability).

Several aspects of the flow are estimated as functions, inside a moving time window. If some of these functions become very large during a short time scale, a TIP is identified for a duration of τ years. The seismic area is divided into several overlapping parts. Their diameter L is expressed in terrestrial degrees at the equator as a function of M_0:

$$L = \exp.(M_0 - 5.6) + 1 \qquad (2.53)$$

These functions are determined independently for each part. They are related to the seismic history of the zone, and are established from a series of earthquakes of the type studied here (the main earthquake of magnitude $\geq M_0$). Time and space windows are used to separate the after-shocks.

Each main earthquake in the series is defined by a vector with 6 components $\{t_i, \phi_i, \lambda_i, h_i, M_i, B_i(e)\}$ where i is the number of occurrence of the earthquake, t the beginning time, ϕ, λ, and h the latitude, longitude and depth, M the magnitude, and B(e) the number of after-shocks that occurred in the first two days after the earthquake.

The intensity of the earthquake flow (i.e. the current level of seismic activity) is represented by the number of main earthquakes which occurred during the time interval s(t-s, t) and possesses a magnitude greater than the lower limit $\underline{M}$. This number is noted $N(t: \underline{M}, s)$.

The deviation of seismic activity from the long-term linear tendency is characterised by the function:

$$L(t: \underline{M}, s_0) = N(t: \underline{M}, t - t_0) - N(t - s: \underline{M}, t - s - t_0) \frac{(t - t_0)}{(t - s - t_0)} \qquad (2.54)$$

where t_0 marks the beginning of recording.

How is the clustering of the main earthquakes in space measured ?

$$S(t: \underline{M}, \overline{M}, s, \alpha, \beta) = \sum 10^{\beta(m_i - \alpha)} \qquad (2.55)$$

is the weighted sum of the main earthquakes during the time interval (t - s, t) and for the magnitude intervals $(\underline{M}, M)$. M_i is the magnitude of earthquake i, lying between $\underline{M}$ and M. Each weighting depends on the magnitude, and is approximately proportional to the source's length. Specifically, one uses $\beta = b/3$, where b is the coefficient of the magnitude-energy relation $\log.E = a + b \times M$. α is a normalisation parameter. The average length of the source is proportional to s/N and the average distance between sources is proportional to $N^{-1/3}$, in the case of a uniform distribution. Their ratio is characteristic from the concentration and can be roughly estimated by the function:

$$Z(t: \underline{M}, M, s, \alpha, \beta) = \frac{S(t: \underline{M}, M, s, \alpha, \beta)}{\left[N(t: \underline{M}, s) - N(t: M, s)\right]^{2/3}} \qquad (2.56)$$

The clustering of earthquakes is described by the recording of after-shocks, for one year before t, for the main earthquakes whose magnitude lies between $\underline{M}$ and M. Therefore:

$$B(t: \underline{M}, M, 1 \text{ year}) = \max. \{B_i(e)\} \qquad (2.57)$$

As the intensity of the earthquake flow varies from one region to the other, the flows are normalised by adjusting the threshold of the magnitude $\underline{M}$ so that the yearly average is identical to a common value. For a series of functions noted N_1, L_1 and Z_1, the constant may be of 10 per year, and for another series of functions noted N_2, L_2 and Z_2, the constant may be of 20 per year. For these 6 functions, the length of the time interval s is 6 years. The upper magnitude limit M is adjusted to M_0. And for the function Z, we use: $M = M_0 - 0.5$; for B, the thresholds are $\underline{M} = M_0 - 2$ and $M = M_0 - 0.2$.

The earthquake flow at the instant t is therefore represented by a vector of seven functions, evaluated at the time before t.

The problem is to decide if t marks the beginning of a TIP corresponding to a large earthquake with $M \geq M_0$. This is the case if very high values of the functions are clustered in a time interval short enough.

To predict a TIP beginning at instant t, we must check that for the last three years (including t and excluding t - 3): (1) each group $\{N_1; N_2\}$, $\{L_1; L_2\}$, $\{Z_1; Z_2\}$ and $\{B\}$ contains functions with very high values; (2) at least six of the functions N_1, N_2, L_1, L_2, Z_1, Z_2 or B have very high values.

In other words, at least six of the seven groups of six estimates (one group for each function) must have extremely high values and one of them corresponds to B. The term "high value" or "very high value" concretely means values of the order of $Q = 10$ for all functions except B and $Q = 25$ for B.

The vector is computed for discrete values of time T_j, with a step of half a year. The interval $(T_j, T_j + \tau)$ is considered to be a TIP if the conditions presented above are true at T_{j-1} and T_j. The duration τ of the TIP was chosen as 5 years. One considers that at least 12 years of data since t_0 are necessary for a valid determination of the functions. It is then possible to analyse the time intervals following $t_0 + 12$.

The research of Keilis-Borok and Kossobokov (1990) has been applied to several regions and on the world's seismicity, in all 19 regions with magnitudes M_0 varying between 4.9 (Koyna Dam) and 8.0 (Central America and global seismicity). Table 2.2 sums up the success rates of the method (Keilis-Borok, 1990; Kossobokov, 1990).

Table 2.2: Success rates of the M8 method.

	Strong earthquakes		TIPs	
	Total	in TIPs	Total number	Number of successes
Regions 1-19	44	39 (89%)	72	49
Regions 2-15	28	25 (89%)	38	28

Figure 2.34 shows three tests of the seismicity of California. Out of three large earthquakes, two were predicted *a posteriori*, and one of them in advance (Loma Prieta, for October 1989 instead of 1990) (Kerr, 1991).

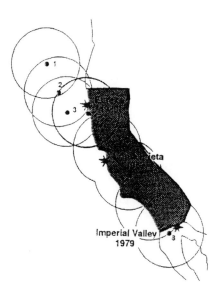

Figure 2.34. *A posteriori* results of three tests of the M8 algorithm on the seismicity of California. Out of three major earthquakes, two TIPs were verified.

2.7.3 The CN Algorithm

This algorithm is very close to the M8 algorithm, and is based on a larger number of characteristics of the earthquake flow: seismic activity level, its time variations, the clustering of earthquakes in time and space, their concentration in space, and their large-scale interaction. The CN algorithm is presented in detail in an article by Keilis-Borok and Rotwain (1990).

An earthquake is still defined as important if its magnitude M is larger than a given threshold magnitude M_0. The identification of a TIP is based on the characteristics of the earthquake flow for a larger class of magnitudes. These characteristics are time functions defined over series of earthquakes, in a given region, inside a moving time window. These functions are normalised so that they can be uniformly applied to regions with other levels of seismicity.

Let us consider a series of major and precursor earthquakes in a given region. t_i is the start hour and M_i the magnitude of the i-th earthquake, $t_{i+1} > t_i$. The flow functions are defined on a moving window $t - s \leq t_i \leq t$.

- The level of seismic activity $N(t \mid \underline{M}, s)$ is defined as the number of major earthquakes of magnitude $M \geq \underline{M}$.

$$\sum (t \mid \underline{M}; \overline{M}; s; \alpha; \beta) = \sum_i 10^{\beta(m_i - \alpha)} \qquad (2.58)$$

is the number of major earthquakes weighed with M_i. This sum is made on the major earthquakes (or main shocks) such that:

$$\overline{M} \geq M_i \geq \underline{M} \qquad (2.59)$$

β and α are the same parameters that were previously defined for the M8 algorithm.

- The rest function is defined as the activity deficit:

$$q(t \mid \underline{M}, s, u) = \int_{-\infty}^{+\infty} POS\left[n(\underline{M})s - N(r \mid; \underline{M}, s)\right] dt \qquad (2.60)$$

POS means that only the positive values between the brackets are considered. Among the different types of rest, this function allows us to identify the seismic gaps.

A second type of rest function can be defined as the value of $N(t)$'s last minimum.

$$Q(t \mid \underline{M}, s) = \left[N(t_1 \mid \underline{M}, s) - N(t_2 \mid \underline{M}, s)\right] - \left[N(t \mid \underline{M}, s) - N(t_2 \mid \underline{M}, s)\right] \qquad (2.61)$$

This function Q only concerns the last fifteen years and the time interval is therefore $(t - 15, t)$, t_1 is the time of the most recent maximum of the function $N(t)$ and t_2 is the time of a minimum of $N(t)$ between t_1 and t.

- The temporal variation of seismicity is:

$$L(t \mid \underline{M}, s) = N(t \mid \underline{M}, t - t_0) - N(t - s \mid \underline{M}, t - s - t_0) \frac{t - t_0}{t - s - t_0} \qquad (2.62)$$

This corresponds to the deviation from the long-term tendency. t_0 is the time when recording starts. The second term of the equation is the linear extrapolation of N from time $(t - s)$ to time t.

$$K(t \mid \underline{M}, s) = N(t \mid \underline{M}, t - s) - N(t - s \mid \underline{M}, s) \qquad (2.63)$$

is the difference between the number of major earthquakes for two successive time intervals.

These three types of functions were presented in detail to show the idea behind their respective definitions. The article of Keilis-Borok and Rotwain (1990) then presents the functions of: spatial concentration, seismic clustering, spatial activity contrast, large-scale seismic interaction, etc. 18 to 20 functions are thus defined for use in the CN algorithm, and more can be added at will.

Ch. 2] Algorithms M8 and CN 135

Let us look now at the pattern recognition side of the problem. In a large area divided in several regions, the earthquake flow in region K can be represented by the vector:

$$\mathbf{P}(K,t) = \left[P_1(K,t); \ldots ; P_m(K,t)\right] \qquad (2.64)$$

where $\mathbf{P}(K,t)$ is one of the functions defined above. The problem can then be set out as follows.

Knowing $\mathbf{P}(K,t)$, the instant t must be recognised as a TIP, or not as a TIP, i.e. we must decide whether the probability of a strong earthquake ($M \geq M_0$) in the region K during the time interval $(t, t + \tau)$ has increased or not. This is a typical pattern recognition problem with a small number of samples (here, the functions).

The CN algorithm uses the specific approach laid out by Gelfand et al. (1976). Time is divided into intervals of two types (Figure 2.35): $D(t_1)$ before each strong earthquake, N for the rest of the time. Inside some parts of the D intervals, incoming strong earthquakes create abnormal earthquake flows, and therefore D is superposed on the real TIP. The aim is to recognise the combinations of "region, time" (X,t) described by the vector $\mathbf{P}(K, t)$. The rule for recognising the TIPs stems from the original samples in intervals D and N.

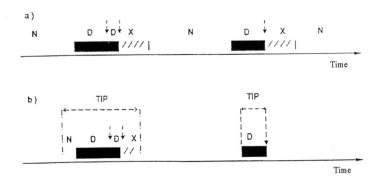

Figure 2.35. The CN algorithm is constructed using the approach of Gelfand et al. (1976): interval (D, t_1) before each strong earthquake, N for the rest of the time.

Let us see the application to earthquakes in California and Nevada with $M \geq 6.4$, between 1938-1988. The original samples consist of:

- objects D: 15 sub-classes corresponding to periods of 2 years before each major earthquake.

- objects N: 24 moments of period N 3 years after each major earthquake.

- for each of the functions defined earlier, we estimate which of its values (large, small, average) is the most typical for D or for N.

- The diagnosis of TIPs comes later: each region, at a given time has a number $n_D(t)$ of features characteristic from D and a number $n_N(t)$ of features characteristic from N. The pattern recognition algorithm is here based on the difference:

$$\Delta(t) = n_D(t) - n_N(t) \tag{2.65}$$

  The difference is computed for each region with a time step of 2 months.

- Because of the region's seismo-tectonic history, $\Delta(t) \geq 5$ during a year after each major earthquake (there is only one exception).

  The TIP will start at the time $\Delta(t) \geq 5$ and last 1 year therefrom, or until the beginning of the strong earthquake if it occurs less than a year after the start of the TIP.

Figure 2.36 shows the results obtained for the 1938-1988 records in the southern and northern regions of California. TIPs correspond in average to 24% of the time, and precede 23 of 29 major earthquakes, and 17 out of 25 TIPs are immediately followed by a strong earthquake. The success/failure ratio is therefore relatively good.

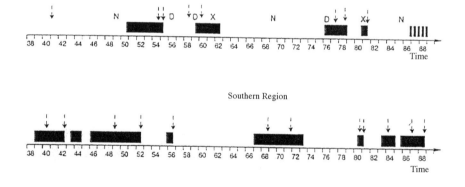

Figure 2.36. Results obtained with the CN algorithm. The data come from the 1938-1988 records in the southern and northern regions of California. The arrows point to the occurrences of strong earthquakes.

2.8 GEOMAGNETISM

The variations of the Earth's magnetic field in time and in space have been the subject of a very large number of studies. Among the syntheses available, Courtillot and Le Mouël (1988) present a graph of energy spectra for variation periods on twelve orders of magnitude, summing up the works of Filloux (1980), Barton (1982, 1983) and Roberts (1986) (Figure 2.37).

The appearance of these spectra may lead to the conclusion that the variations observed follow a power law. They are indeed very close to the results obtained in geomorphology and in the study of seafloor profiles (see Section 2.1.2), for which the fractal dimension was related to the slope of the energy spectrum.

But variations of the Earth's magnetic field and specifically the time series of field inversions, have long been admitted to follow an exponential distribution. McFadden (1984) likens it to a distribution Γ. Inversions should therefore occur randomly. The application of the tests described in this chapter show that this is only an approximation, and that determinism can be brought into evidence for some time domains of the field inversions.

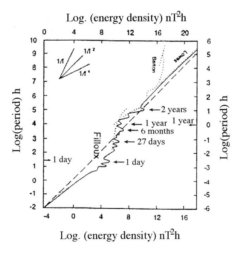

Figure 2.37. Energy spectra for the periods of variation of the Earth's magnetic field. This graph spans twelve orders of magnitude, using data from Filloux (1980), Roberts (1986) and Barton (1982, 1983). From Courtillot and Le Mouël (1988).

2.8.1 Inversions of the Earth's Magnetic Field

Our study will concentrate on two objects. The first will be the physical model of the two-disk dynamo (Rikitake, 1958), which accounts simply for a possible mechanism of field inversion. This model was recently studied as an unstable dynamic system by

Ershov et al. (1989) (see also Turcotte, 1992), using the techniques of non-linear dynamics (mainly search and description of the system's attractor). The second object will be the Cox scale (Cox, 1964), for an empirical study of inversions using the correlation function (Dubois and Pambrun, 1990).

Generalised model of the two-disk dynamo (Ershov et al., 1989)

Ershov et al. (1989) use the model of a dynamo made out of two disks with friction, shown in Figure 2.38. Rikitake's dynamo is a system where the two disks are connected. Their respective axes are submitted to identical torques. Adding viscous friction, which reduces the angular moment of the disks, the system can be described by the following system of equations:

$$\begin{cases} dx_1/dt = -\mu x_1 + x_2 x_3 \\ dx_2/dt = -\mu x_2 + x_1 x_4 \\ dx_3/dt = 1 - x_1 x_2 - v_1 x_3 \\ dx_4/dt = 1 - x_1 x_2 - v_2 x_4 \end{cases} \qquad (2.66)$$

These variables are dimensionless. x_1 and x_2 are the electric currents in the disks, x_3 and x_4 are the angular speeds, μ is the ohmic dissipation coefficient (identical for each circuit), and v_1 and v_2 are the viscous friction coefficients (they can be different for each disk).

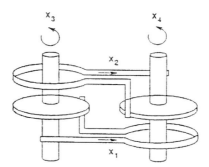

Figure 2.38. Model of the two-disk dynamo (Ershov et al., 1989). This model is the same as Rikitake's (1958). x_1 and x_2 are the dimensionless values of the electric currents in the disks, x_3 and x_4 are proportional to the disks' angular speeds

Several models were built along these lines (Allan, 1962). Cook (1972) studied an identical model with different moments on each disk, but with $v_1 = v_2$.

When $v_1 = v_2 = 0$, the last two equations can be reduced, and the system becomes Rikitake's model again with $x_3 - x_4 = A$ (constant). According to the values of A and μ,

Rikitake's system may go toward a limit cycle in which the $x_i(t)$ are periodic, or toward a strange attractor in which the $x_i(t)$ oscillate irregularly. For example, for values of $\mu = 1$ and $A = 3.75$, the system exhibits a chaotic behaviour (Cook and Roberts, 1979). Using the method of Wolf et al. (1985) (see Section 1.5.6), the Lyapunov exponents are determined to be:

$$\lambda_1 \approx 0.12 \ ; \ \lambda_2 = 0 \ ; \ \lambda_3 = -2.19 \tag{2.67}$$

The existence of positive exponents implies an exponential divergence of the phase trajectories. The dimension of the strange attractor is given by the Kaplan-Yorke formula:

$$d_1 = (2 + \lambda_1) / |\lambda_3| \approx 2.09 \tag{2.68}$$

When the viscosities v_1 and v_2 are different from 0, the system is written as:

$$\begin{cases} dx_1/dt = -\mu x_1 + x_2 x_3 \\ dx_2/dt = -\mu x_2 + x_1 (x_3 - A) \\ dx_3/dt = 1 - x_1 x_2 - v_1 x_3 \end{cases} \tag{2.69}$$

$$dA/dt = -v_2 A - \Delta v x_4 \tag{2.70}$$

where $x_3 = x_4 + A$ and $\Delta v = v_1 - v_2$. The parameter A was constant in Rikitake's system. It varies now with time, and does not tend toward a constant when $t \to \infty$.

Ershov et al. (1989) study the system's attractor on the projections $x_1 x_2$, $x_3 x_4$ and $x_1 x_3$ for several values of the parameters μ, v_1, v_2 (Figure 2.39). For example, in the case of $\mu = 1$, $v_1 \approx 0.004$; $v_2 = 0.002$, the attractor is very thin (cf. the projection on $x_3 x_4$ in Figure 2.39).

The different projections of points representative of the dynamic system in the phase space are varying chaotically from positive to negative values of x_1 and x_2. These two variables are at the origin of the magnetic fields created by the disks. Fields parallel to the axes add up and produce a resulting field with can be positive or negative, depending on the phase trajectory on the attractor. Each time the sign changes, there is inversion. This means the series of inversions along time are chaotic (determinist).

Many numerical tests performed by the authors (cf. Figure 2.39) allow to conclude that:

1. Contrary to Rikitake's dynamo, the friction dynamo "forgets" the initial conditions.
2. For all values of the parameters, there exists a stable equilibrium.
3. The attractors' dimension is 3+ε, with 0 < ε < 0.2, while it is 2+ε for Rikitake's model.
4. Asymmetric cycles appear, but the attractors are quasi-symmetric.
5. When ohmic resistivity decreases, the path toward chaos shows bifurcations with doubling of the period. This happens as well when friction decreases.

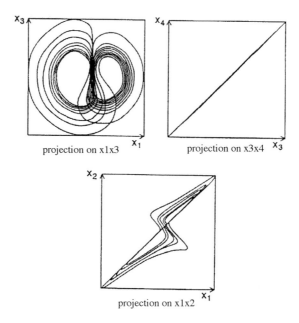

Figure 2.39. Attractor of the dynamic system associated to the model of Erschov et al. (1989), for $\mu = 1$, $v_1 \approx 0.004$; $v_2 = 0.002$. The three graphs represent the respective projections on x_1x_3, x_3x_4 and x_1x_2.

Study of inversions: Cox scale

This study is a summary of the results obtained by Dubois and Pambrun (1990) when working on the series of inversions of the Earth's magnetic field (Cox scale). A first part of the study looked at the distribution of time intervals between two successive inversions. 296 periods were investigated (148 direct and 148 inverse orientations of the magnetic field), with durations between 0.01 and 35.25 Ma.

Figure 2.40 shows the distribution of these periods every 5 Ma, from -165 Ma to the present period. It also shows the cumulated values of inverse and direct periods of durations $\geq$ t, on three graphs: N as a function of t, log.N as a function of t and log.N as a function of log.t.

The semi-log graph shows a quasi-linear slope of -3 between 0 and 0.55 Ma. It is used to deduce that the distribution is exponential ($N = N_0\, e^{-3t}$) in the interval [0; 0.55]. This result is in agreement with the conclusions of McFadden (1984), who likens it to a distribution Γ. This distribution applies for 200 of the 296 events. For t > 0.55 Ma, the distribution then moves away from the regression line. But the log-log graph shows a quasi-linear slope of -1.6 between 0.55 and 2 Ma. A law of the type $N = N_0\, e^{-1.6t}$ can be deduced for the remaining 90 events.

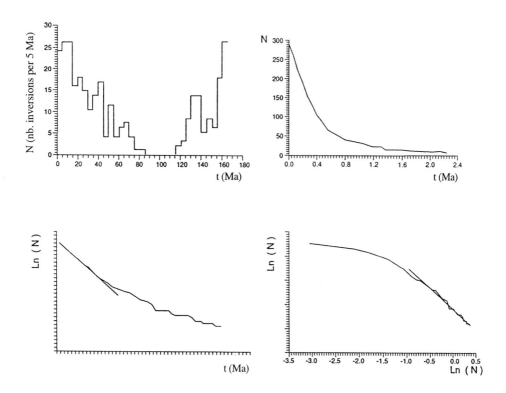

Figure 2.40. Inversions of the Earth's magnetic field. These four graphs represent (from left to right and top to bottom): the frequency of inversions every 5 Ma, from -165 Ma to now, and the distribution of cumulated values of durations $\geq t$, with N as a function of t, log.N as a function of t and log.N as a function of log.t. From Dubois and Pambrun, 1990.

The random Poissonian character of the first distribution is compatible with it, but maybe not for the second distribution. Investigations performed on dynamic systems (Arnéodo and Sornette, 1986) demonstrated that these two types of distribution may both be the expressions of chaotic determinist systems with more or fewer degrees of freedom.

This led us therefore to test the determinism of the system generating the series, with the correlation function of Grassberger and Procaccia (1983). This correlation function is applied to a pseudo-phase space constructed with the durations between inversions X and their occurrence number. In a three-dimensional space, for example, the point representative of occurrence i will be plotted with the values x_i, x_{i+1}, x_{i+2}.

The results are presented in Figure 2.41. There is no saturation of the value of the slope of log.C(r)/log.r (see Section 1.5.4), when the dimension of the phase space increases (periods [0 to -168 Ma] and [0 to -23 Ma]). However, between -23 and -168 Ma, the saturation indicates that the system has an attractor of dimension slightly greater than 2. This reminds of the value obtained for the simple Rikitake model.

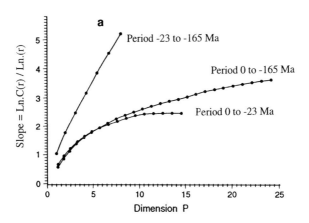

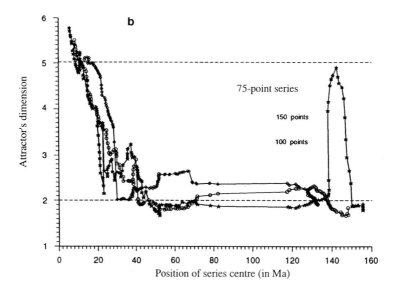

Figure 2.41. Application of the method of Grassberger and Procaccia (1983) to the series of inversions of the Earth's magnetic field (Cox scale). The slope of the correlation function is represented as a function of the dimension of the phase space for the whole dataset (top) and when using a moving window (bottom). From Dubois and Pambrun, 1990.

Chaos therefore appears to be determinist in this period, which confirms the observation on the distribution law for periods greater than 0.55 Ma. When introducing the inversions separated by short periods, in recent times, it seems that the system is perturbed by white noise.

One of the main problems encountered in this kind of study is the small number of data points accessible (around 300), and the error bars on the dates of the oldest inversions. The developments of palaeomagnetism lead to constant progress in the dating of rocks and sediments, and in the possibility of going further than the -168 Ma of the Cox scale. It is therefore reasonable to hope for improved studies in the near future, using continuous series up to 300 or 350 Ma (cf. Besse and Courtillot (1991) and Gallet et al. (1992)).

2.8.2 Temporal Variations of the Magnetic Field Vector

Continuous recording of the components of the magnetic field vector only started at the beginning of the nineteenth century. Good quality data goes back at most a century. We have investigated the records made at the Observatoire National of Chambon-la-Forêt (France) between 1883 and 1992, which are amongst the most reliable.

We shall still use the correlation function, which is well suited here for time series sampled regularly. The time series cover a time domain from one minute to 110 years, i.e. a bit more than 7 orders of magnitude. We limited ourselves to the search for a possible attractor on hourly, daily, monthly and yearly series of the mean values of the field's component, of its module F and of its declination D. The records of Chambon-la-Forêt show that (Dubois et al., in prep.):

1. These series are noisy, except for the series of averaged yearly values, which saturates at $v = 1.3$ (v is the slope of the line log.C(r)/log.r, i.e. the attractor's dimension).
2. The values of v decrease from 3.5 (for a dimension of 10 in the phase space), for hourly series, down to 1.3 for the yearly series.
3. Computation with a moving window shows that the dynamic systems generating these series are not stable. This means that the dimension of their attractor varies with time (this may cause the observed noise).
4. These preliminary observations should be extended to other series recorded at other places. They show that the number of degrees of freedom of the dynamic systems ruling the magnetic field variations are higher when the study interval is shorter. This can be explained by the fact that these variations are the sum of various internal and external processes: storms, daily and seasonal variations, etc. These variations are filtered better when the values are averaged on longer time scales. The number of degrees of freedom for the system corresponding to hourly variations is necessarily higher than the one for daily variations, itself higher than the one for monthly variations, etc.

Study of secular variations

Researchers interested solely in the internal causes of variations of the Earth's magnetic field now have the possibility of using palaeomagnetism, as for the study of inversion series. Since the recent studies of cores from lacustrine sediments (Thouveny, 1991) or marine sediments (Tauxe and Wu, 1990), access is possible to time series down to 120,000 years ago. For lacustrine sediments, the sampling interval is around 200 years.

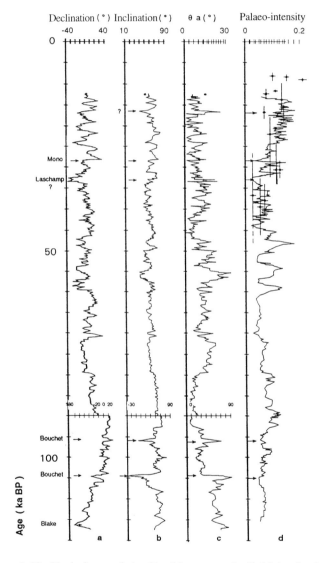

Figure 2.42. Variations of the Earth's magnetic field in the last 120,000 years. The four graphs correspond to: (a) the vector's declination; (b) the vector's inclination; (c) the vector's deviation; (d) the relative palaeo-intensity. From Thouveny, 1991.

The measures made in observatories and these measures are nearly continuous in time. And the lower limit of 120,000 years is very close to the limit of the periods of direct or inverse polarity analysed in the previous section.

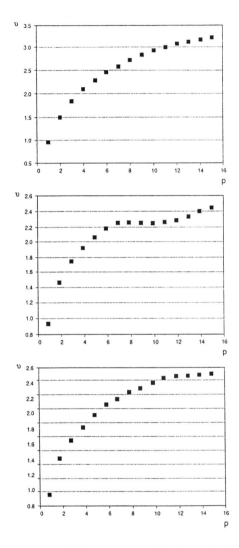

Figure 2.43. Dimensions of the attractors for the variations of the Earth's magnetic field. The variations of the slope are measured as a function of the dimension of the phase space (method of Grassberger and Procaccia). Each series contains 498 values of D, I, F measured on sediments from Lake Bouchet. There is a good saturation of the inclination for a value of 2.3.

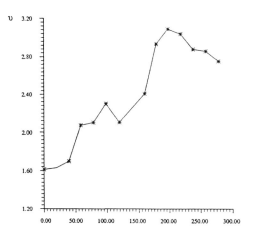
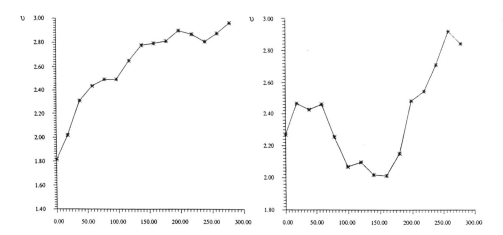

Figure 2.44. Moving-window analysis of the attractor's dimension along a core from Lake Bouchet. The dimension of the phase space has been fixed at 7 for the three components (Atten and Caputo, 1987).

The time domain of field variations is therefore covered from one second (rapid fluctuations recorded continuously) to 500 Ma, i.e. more than 16 orders of magnitude. Palaeomagnetic variations recorded in the sediments from Lake Bouchet (Thouveny, 1991) are represented on Figure 2.42. The variations of declination, inclination and palaeo-intensity appear to be chaotic. Analysis with the method of Grassberger and Procaccia leads to the graphs presented in Figure 2.43, plotted for 498 values in each of the three series. The main observations are:

1. There is saturation, or a tendency toward saturation, for dimensions of the phase space close to 7 - 9. The values of v are: 3 for D, 2.4 for I and 2.6 for F.
2. The declination seems the most noisy of the three variables, which seems logical because of the techniques used to place the cores in space.
3. The values of v range between 2 and 3. This implies the existence of an attractor ruling the behaviour of a determinist chaotic system with three degrees of freedom.

The evolution of the system with time was studied using a moving window of 200 points (Figure 2.44), moved with a step of 20 points. Even if the system is not perfectly stationary, the attractor's dimension varies between 2 and 3.

Because of its small dimension (smaller than 3), we have represented the attractor in its phase space with the axes $X(t), X(t+\tau), X(t+2\tau)$ where X is the value of I and $\tau = 208$ years (Figure 2.45). The main observations are:

1. All phase trajectories are inside a finite volume (cocoon-shaped).
2. They do not fill this volume entirely, which means the attractor's dimension is smaller than 3.

The third stage (Hongre et al., in prep.) consists in using a theoretical model of the magnetic field and its coefficients g_n^m and h_n^m, to compute the theoretical variations sampled every 200 years, and determine on this theoretical series the slope of the correlation function. The computation deals with:

$$I(t) = \arctan \frac{Z(t)}{\sqrt{X^2(t)+Y^2(t)}} \qquad (2.71)$$

where:

$$X(t) = \sum_n \sum_{m=0}^n (g_n^m(t)\alpha_m + \beta_m h_n^m(t))\delta_n^m) \qquad (2.72)$$

$$Y(t) = \sum_n \sum_{m=0}^n (g_n^m(t)\beta_m - \alpha_m h_n^m(t) \gamma_n^m/\sin\theta) \qquad (2.73)$$

$$Z(t) = \sum_n \sum_{m=0}^n (g_n^m(t)\alpha_m + \beta_m h_n^m(t))(n+1)\gamma_n^m) \qquad (2.74)$$

and where $n = 1, ..., \infty$ and $m = 0, ..., n$; θ and ϕ are the coordinates of the measure point (here, in Lake Bouchet), and with the Legendre polynomials:

$$\alpha_m = \cos.m\phi \; ; \; P_n^m(\cos\theta) = \gamma_n^m \qquad (2.75)$$

$$\beta_m = \sin.m\phi \; ; \; \partial P_n^m(\cos\theta) / \partial\theta = \delta_n^m \qquad (2.76)$$

In studies conducted later, a generating function g(t) has been introduced in the 24 coefficients g_n^m amd h_n^m (we stopped at n = 4), keeping in mind the constraints given by Hulot et al. (1991): random variations and exponential decreasing of their amplitude when n increases. This generator was used with three sinusoidal terms of periods 220, 550 and 1200 years, and random phases. The resulting variations of the theoretical field are very interesting, and result from a system with an attractor of dimension 2 < D < 3.

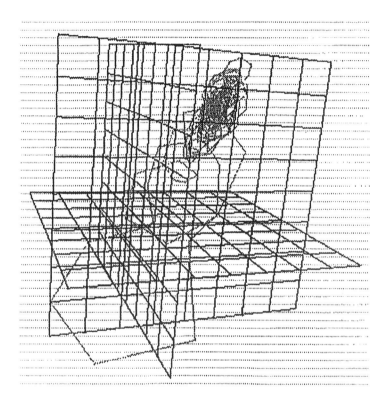

Figure 2.45. Perspective representation of the attractor of the magnetic field variations. It has been tilted in a three-dimensional space (the attractor's dimension being 2.3). The algorithm was applied on the CM-2 supercomputer by P. Stoclet.

3

Perspectives

> *The world is my representation*
> Schopenhauer, 1844

3.1 GENERAL POINTS

> *The world is everything that occurs*
> L. Wittgenstein, Tractatus
> logico-philosophicus, 1918

The previous chapters showed that natural processes often result from the interaction of dynamic systems with a relatively high number of degrees of freedom (usually more than 2), which give them a disorderly appearance. These processes are ideally suited for non-linear approaches.

But many domains have still not been examined in this optic. The major works dealt with geomorphology and matters related to rock fragmentation and fracturation, rock mechanics, tectonics, seismology and volcanology. However, the boundaries between these disciplines are as promising, for example with the study of hydrology in relation with topography and geomorphology, hydrogeology and fluid circulation in porous media, percolation and its application to hydrocarbon reservoirs and geothermalism, or the propagation of pollution fronts.

The possibility of computing with vectors with n components broadens considerably what was usually reduced by the differential equations representing dynamic systems to the single time variable, and usually encountered problems with non-integrability.

This broadening of horizons is a direct consequence of the study of the dynamic systems which interact on our planet. Fluid circulation allows to link different disciplines previously separated, such as hydrology, hydrogeology, physics and mechanics of fissured environments, geomagnetism, electromagnetic induction and electric currents, electrofiltration, etc. This shows the unifying character that results from the introduction of new paradigms, in the meaning of T. Kuhn. This richness of approaches can also be found inside the discipline itself. Seismology is one example among many. Smalley et al. (1987) conclude their study of seismicity along one dimension, time (see Section 2.3), by a fractal analysis and the exposition of their project: *"the initial goal of our study was to present a fractal analysis of seismic clustering. But earthquakes consitute events with five dimensions: time, three spatial (Euclidean) dimensions, and magnitude. Clustering could have been as well studied in each of these dimensions. One of the aims of this type of investigation would be to determine if the change in fractal dimension is associated to the clustering of events, and would then be a precursor signal"*.

Russian seismologists led their research along this line (Keilis-Borok, 1986, 1988, 1990). The resulting M8 and CN algorithms use many parameters processed as vectors with up to 20 dimensions, and representing the seismic flow (Keilis-Borok and Rotwain, 1980).

Some fields of Earth sciences still seem far from a possible analysis with the tools from non-linear dynamics: petrology, mineralogy, sedimentology, stratigraphy, palaeontology. The last domain has however been approached at least once (Chaline et al., 1992; Dubois et al., 1992). Studying the evolution of fossil species showed that the series of speciations and extinctions of a particular family (the European *Arvicolidae*), in the last 5 Ma, was determinist. This result clearly opposes the traditional concept of random evolution, and is but a first stage, as the number of degrees of freedom of the dynamic systems governing biological evolution still needs to be determined. However, this behaviour with time should not be surprising if examined in conjunction with results obtained in non-linear, short-timescale studies in biology (Thom, 1972; MacKey and Glass, 1977; Sugihara and May, 1990).

It would be possible to list all the disciplines that will, at one time or another, benefit from non-linear studies. We have decided instead to show the advancements allowed by the transition from similarity dimensions to information dimensions, then to correlation dimensions, and finally generalised dimension (Section 3.2). The introduction of multifractals has allowed us to look at measures on the fractal support, and the wavelet transform has proved to be a "mathematical microscope" for studying very complex datasets (Section 3.3). In a probabilist approach, we shall also insist on the concept of self-organised criticality (Section 3.4); there too, new tools are being developed to investigate chaos, intermittent series and catastrophes. Most datasets are recorded for discrete intervals of time. The problem of short series is currently one of the most important, and is granted a section (Section 3.5).

3.2 CHAOTIC TIME SERIES

Time series are often the basic data on which geophysical studies are conducted. The previous chapters showed that, when the series seem chaotic, algorithms existed to analyse them. Examples are the correlation function algorithm (Grassberger and Procaccia, 1983; Malraison et al., 1983) or the algorithms for the computation of the Lyapunov exponents (Eckman et al., 1986; Wolf et al., 1985). They aim at determining the geometric invariants (in the phase space) and dynamic invariants of an underlying strange attractor, such as the correlation dimension, the information dimension, or the Lyapunov exponents.

If this has been performed correctly, it is possible to say that, in a certain way, the chaos has been quantified and a model was proposed. As in any search for a model, this poses the question of data inversion. To the best of our knowledge, the traditional methodology of data inversion (cf. Backus and Gilbert, 1967; Dehlinger, 1978; Tarantola and Valette, 1982; Menke, 1989) has only be applied to a few cases (Farmer and Sidorowich, 1987, 1988; Casdagli, 1989; Sujihara and May, 1990).

In a theoretical study, Casdagli (1989) underlines that the amount of data necessary for these algorithms can be prohibitively high, and that it may be difficult to assess if the proper amount has been reached. This can lead to the misdiagnosis of time series as chaotic determinist. But, even if the algorithms have been successfully applied, the invariants computed can show only a limited practical value. It is possible to deduce from these tests that a time series is determinist, and that it is possible, in principle, to build a predictive model. The largest Lyapunov exponent hints to the limit of predictability, and the correlation or information dimension is an indication of the model's complexity. However, no idea is produced concerning the way to construct the predictive model itself.

Following a different approach, Casdagli (1989) suggests that a predictive model can be constructed straight from the time series. We shall develop his method.

He first deals with the inverse problem.

$f: \mathcal{R}^n \to \mathcal{R}^n$ is a smooth map with a strange attractor α, and $x_n = f^n(x_0)$, $1 \leq n < \infty$ is a typical series of iterations of f over α.

The inverse problem consists in building a smooth map $f_\infty: \mathcal{R}^n \to \mathcal{R}^n$ as a function of x_n so that:

$$x_{n+1} = f_\infty(x_n), \quad 1 \leq n < \infty \tag{3.1}$$

This inverse problem has a unique solution, as the x_n are dense over α and f_∞ is smooth. Therefore:

$$f_\infty |_\alpha = f |_\alpha \tag{3.2}$$

There are no restrictions on the behaviour of f_∞ outside the attractor α. However, f_∞ is perfect predictive model of f over α, and the theory of the inverse problem is trivial.

The number of data points is never infinite, and a more realistic inverse problem consists in building the map $f_n: \mathcal{R}^n \to \mathcal{R}^n$ with a finite number of iterations x_n, $1 \leq n \leq N$, for which:

$$x_{n+1} = f_n(x_n), \quad 1 \leq n \leq N \tag{3.3}$$

In this case, the problem has many solutions and this can be demonstrated easily (see Casdagli, 1989). For a given technique of interpolation, f_n is well defined and may be considered to be a predictor.

To quantify the quality of the predictor f_N of f, the error $\sigma(f_N)$ on the predictor is defined by:

$$\sigma^2(f_N) = \lim_{m \to \infty} \frac{1}{M} \sum_{n=N}^{N+M-1} [x_{n+1} - f_n(x_n)]^2 / \text{Var} \tag{3.4}$$

whre Var is a normalisation factor (i.e. a variance), defined by:

$$\text{Var} = \lim_{m \to \infty} \frac{1}{M} \sum_{n=N}^{N+M-1} x_m \left(\lim_{m \to \infty} \frac{1}{M} \sum_{n=N}^{N+M-1} x_m \right)^2 \tag{3.5}$$

The fundamental theoretical problem with the inverse method lies in the establishment of the convergence characteristics for $\sigma(f_N)$, which are called scaling laws. The predictor f_N is said to be of order γ if:

$$\sigma^2(f_N) = O(N^{-\gamma/D}) \tag{3.6}$$

when $N \to \infty$, andwhere D is the dimension of the attractor α.

For $\gamma = 2$ (Farmer and Siderowich, 1987), $\sigma(f_N) = O(N^{-2/D})$. This is an approximation, but its validation on known applications has given satisfactory predictors.

Casdagli's method is used practically with local predictors and radial-function predictors. We shall not delve further into the numerical approximation techniques reviewed by Casdagli (1987), and their respective defaults and advantages. Instead, we shall look at polynomial techniques.

Functions of coordinates $\pi_i f_N: \mathcal{R}^n \to \mathcal{R}^n$, $i = 1, ..., m$ are chosen from standard functions. For example, a polynomial predictor $\pi_i f_N$ is chosen as a polynomial of m variables of degrees d or less.

The $\binom{m+d}{m} = \frac{(m+d)!}{m!\,d!}$ parameters possible are chosen so as to minimise:

$$\sum_{n=1}^{N-1} (\pi_i x_{n+1} - \pi_i f_N(x_n))^2 \tag{3.7}$$

This resolves into a linear, least-squares problem. This method yields an optimal polynomial predictor of degree d.

Short-term predictions

The main objective of Casdagli's approach is to predict the immediate evolution of the series observed. This comes down to building a short-term predictor $f_N^{(i)}$ from the points x_n, $1 \leq n \leq N$, for which $x_{n+1} = f_N^{(i)}(x_n)$, $1 \leq n \leq N -i$.

We can use an iterative method, which consists in building the predictor with a step f_N, and consider that the i-th iterate of f_N is:

$$f_N^{(i)} = (f_N)^i \tag{3.8}$$

Along successive iterations, the errors will increase at the rate fixed by the highest of f's Lyapunov exponents λ_{max}, and the error $\sigma(f_N^{(i)})$ on the predictor $f_N^{(i)}$ will be defined by:

$$\sigma(f_N^{(i)}) = e^{i \lambda_{max}} \cdot \sigma(f_N) \tag{3.9}$$

Ikeda's map is defined as $f(x,y) = (1 - \mu (x \cos.t - y \sin.t), \mu \sin.t + y \cos.t)$, with the parameters $\mu = 0.7$ and $t = 0.4 - 6.0 / (1+x^2+y^2)$. With these conditions, Ikeda's map exhibits a strange attractor (Ikeda, 1979).

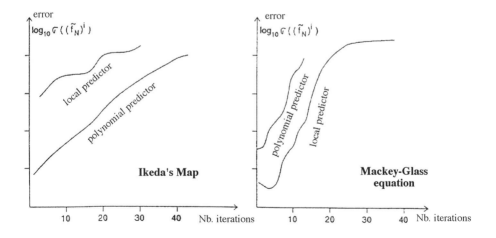

Figure 3.1. Error growth for local and polynomial predictors. These two graphs show how the error increases with i, for Ikeda's map ($\mu = 0.7$) and for the equation of Mackey-Glass. From Casdagli, 1989.

Casdagli (1987) applied this method of short-term prediction onto Ikeda's map and the model of Mackey-Glass (1977). Figure 3.1 shows the errors on the predictors, growing when i increases, for polynomial and local predictors $f_N^{(i)}$.

Sujihara and May (1990) are also investigating the short-term prediction of time series, and they try to distinguish the (random) measure error from the determinist chaos governing the series. They use predicting functions of the same type than those proposed by Casdigli (1987). Their works focus on biological processes, but the analogy with the noisy series studied by geophysicists is striking.

With these methods in mind, it is theoretically possible to start from any time series and predict the next iterations very early in time (1 < i < 20), and the future evolution of the series. The interest for eruptive or seismic series is immediate. In theory, scientists aim at determining the first next iterate from the last one known, and know the time of occurrence of the next eruption or the next earthquake. But the whole difficulty resides in the search for f_N. And let us stress that the method is only usable if the series is generated by a determinist chaotic system.

3.3 FRACTALS, MULTIFRACTALS AND WAVELET TRANSFORMS

The previous chapters have shown the constant evolution of the description of disordered environments. First with the concept of dimension, and the transition from the similarity (or Mandelbrot) dimension D_0, which characterises the fractal frame of a set, to the information dimension D_1, which contains a supplement of information (the number of structures inside the cube of side ε), and to the correlation dimension D_2, which gives details about the geometry (in the phases' space) of the system's attractor. According to Sugihara and May (1990), the dynamic system generating the set studied provides the "active variables" of the underlying attractor of a given chaotic time series. Finally, we saw the concept of generalised dimension D_n ($2 < n < \infty$)

Multifractals constituted an important step forward, introducing the notion of measure or probability of a fractal frame, but even if the amount of data points allows to plot the singularity spectrum, these points cannot be located geometrically, only statistically. This gap is now bridged with wavelet transforms, which act as a "mathematical microscope" and locate the points of clustering of the "mass or probability" (in a broad sense).

Figure 3.2 sums up these three successive stages for a triadic Cantor set. It is evident that fractal objects will benefit from being studied or re-studied using the most recent conceptual advances.

3.4 SELF-ORGANISED CRITICALITY (SOC)

Continuously unstable systems can be readjusted by catastrophic events, of varying amplitude, which occur at all scales and follow power-laws. This observation was a major discovery.

Self-Organised Criticality

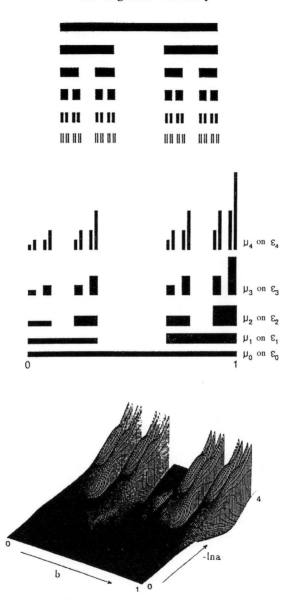

Figure 3.2. The three successive stages of analysis of a triadic Cantor set: (top) the model itself; (b) the multifractal analysis of the model; (c) the wavelet transform. From Mandelbrot (1975), Falconer (1990) and Arnéodo et al. (1988).

First suggested by Bak, Tang and Wiesenfeld (1987), another idea is that there is a close link between scale invariants in the time and space domains. This was also brought into evidence by Sornette and Sornette (1989), who observe that the SOC concept implies that earthquakes result from the reorganisation of the lithosphere at spatial and temporal levels. Accordingly, this concept casts a doubt on the validity of seismic models which only consider a single fault with "bumps" coupled only to their closest neighbours. However, the SOC theory implies that stress and deformation fields are coupled on large scales, and therefore that movements along a fault should be compatible with the other deformations around the fault.

Most authors recognise that the conditions of SOC are not clearly understood yet, and that they should be the object of further studies using statistical physics models as well as observations of the geophysical processes at play (Sornette et al., 1990).

3.5 THE PROBLEM OF SHORT SERIES

Short series very often cause problems during the practical processing of non-linear dynamics datasets, mainly when dealing with dynamic systems. When computing the similarity dimension D_0, using techniques such as Cantor dust or box-counting, the number of values in the series can go down to around 20. This is the limit that Wickman (1966, 1976) selected for his statistical analysis of volcanic eruption series (empirical approach). A detailed study of this limit number still has to be conducted. Apart from the strict definition of D_0 and its error bar, there is the problem of validity of the linearity test, which confirms or infirms the fractal character of the dataset studied. The range of the interval between points should also be considered, and should span at least one order of magnitude (Davy et al., 1990).

Difficulties increase when computing the dimensions of higher orders; the information dimension D_1 and the correlation dimension D_2. In dynamic systems, the correlation dimension is generally computed by applying a correlation function to points in the phase trajectory (method of Grassberger and Procaccia, 1983). The smallest number of points necessary has been studied by Pisarenko and Pisarenko (1991, 1995). They remark that high values of D_2 can be unreliable when the attractor's dimension is estimated on time series of intermediate lengths.

If, for example, $D_2 \geq 5$, the values become suspicious when N is of the order of a few hundreds. Limiting rules have been proposed by Smith and Jordan (1988) and Ruelle (1990), and can be written as:

$$N \gg 10^{D_2/2} \text{ and } N \gg 42^{D_2} \tag{3.10}$$

They show that, when D_2 is close to 2, the series need to contain at least 1,764 values. When $D_2 = 3$, this number becomes 74,088, and is larger than 10^8 when $D_2 > 5$.

Determining the dimension of the attractor ruling the air temperature variations in a particular point (an attractor for which a dimension of 6 or 7 has already been suggested), will necessitate the processing of a series of 5.5×10^9 to 2.3×10^{11} values, i.e. the recording of 0.7 to 26 Ma of hourly temperature values !

This problem has already been mentioned in Section 1.5, about short series of magnetic inversions, and non-stationary dynamic systems (works of Guckenheimer, 1986; Havstad and Ehlers, 1989). Studying geomagnetic inversions, Dubois and Pambrun (1991) empirically show that the dimension of Hénon's attractor (d = 1.26) can be defined at 5% precision with only 100 vectors.

Pisarenko and Pisarenko (1991, 1995) present a statistical method for computing D_2, which frees itself from the increase of N when D_2 increases, and is based on the maximum-likelihood criteria. In the case where there are N points (m-dimensional vectors) x1, ..., xN, if r_0 is the maximum distance between two points and if $r_{ij} = |x_i - x_j|$ is the distance betweeen any pair of points, the estimate of the correlation dimension is:

$$\overline{D_2} = \frac{1}{1/L \sum_{i>j}^{N} w_{ij} \log \frac{r_0}{r_{ij}}} \tag{3.11}$$

with $w_{ij} = H(r_0 - r_{ij})$ and $L = \sum_{i>j}^{N} w_{ij}$.

This method can be applied to series of 200 to 300 points. It also allows us to compute the error bar associated with D_2.

3.6 MASTERING AND CONTROLLING CHAOS

The control of chaos was mentioned at the end of Section 1.7 and will be detailed in Annex A.3. Methods similar to the OGY method (Ott et al., 1990) were shown to allow the maintenance of the phase trajectory on a fixed orbit by acting on one of the system's control parameters. This possibility of control relies on a good knowledge of the system's attractor, and of the control parameters. Several researchers demonstrated this was possible, with physics experiments where a system was brought to a stage highly sensitive to the initial conditions. Mastering chaos is relatively easy for coupled, synchronised, chaotic circuits (Pecora et al., 1990), or for a vibrating strip (Ditto et al., 1990), or for experiments in which the control parameters can be easily modified.

The control of chaos is not feasible yet for natural processes, observed with time series and for which the generating mechanism is very often unknown. It is, however, possible to reply to the questions mentioned above: build a Poincaré section in the phase space of the system, look for stable orbits which will sometimes mix with the orbits of the attractor, and identify passively, without influencing the system, which observable parameters can play the role of control parameter.

At least in a first stage, this research domain will enhance the prediction of the future behaviour of a system, waiting for the day where it will be possible to modify the key parameter controlling the system (i.e. pushing the system from a generic unstability to a stable regime).

4

Illustrations of Non-Linear Dynamics

> *Paradox! Immense Paradox! cried Albert*
> *Not as much as you think, I replied.*
> Goethe, The Sufferings of Young Werther

This chapter will present some illustrations of non-linear dynamics, some of them unusual or recreative. They will call upon fractals, dynamic systems, and self-organised criticality. All these illustrations contributed to the development and application of non-linear dynamics to many disciplines.

4.1 FRACTAL SETS

> *Art is made to disturb,*
> *Science reassures*
> Georges Braque

As mentioned in the introduction, the study of fractal objects showed early on the aesthetic aspect of figures generated with more and more powerful computers (see Colour Plate 1, between pp. 192 and 193). But, before talking of fractal graphics, the observation of fractal objects in the universe should be mentioned with a now traditional example.

4.1.1 The Olbers Paradox

According to Kepler, the distribution of celestial bodies in the universe cannot be uniform, as the night sky would not be dark but uniformly radiant. This paradox is often referred to as the "burning sky" paradox. It was studied again by Olbers in 1823, under the name of Olbers Paradox. Fournier d'Albe (1907) and Charlier (1908) noticed that this paradox disappeared if the celestial bodies were satisfying the relation: $M(R) \approx R^D$, with $D < 2$, M being the mass of celestial bodies at the distance R.

The demonstration of Olbers looks at a star at a distance R from the observer, compared to a star at the distance $R = 1$. Its relative luminosity equals $1/R^2$, and its relative apparent surface is $1/R^2$. The apparent luminosity density is therefore the same for all stars. If the universe is uniform, any direction from the observer will at some point intersect the apparent radius of a star. The apparent luminosity density is therefore the same on the whole sky.

If $D < 2$, a non-zero proportion of directions disappears in the infinity without meeting any star. This is reason enough for the sky to be dark.

There is in this demonstration a strong argument in favour of a fractal distribution of matter in the universe (Mandelbrot, 1975). But we now know that the distribution of galaxies does not correspond to homogeneous fractals, whereas a multifractal approach can describe reasonably well the reality (Jones et al., 1988).

4.1.2 Iterated Function Systems (IFS)

Barnsley (1988) studied and rigorously formalised a series of operations building more and more complex images by transformations in metric spaces, and then in spaces of compact non-empty sets with a Hausdorff metrics (i.e. allowing the use of fractal sets). These transformations can be contraction maps, transformations on real axes (thus on 1 dimension), affine transforms in the Euclidean plane, etc.

Let us first define an affine transform in the Euclidean plane.

A transform $\mathcal{W}: \mathcal{R}^2 \to \mathcal{R}^2$, of the form $w(x_1, x_2) = (ax_1 + bx_2 + e; cx_1 + dx_2 + f)$ and where a, b, c, d, e, f are real numbers, is called an *affine transform* in the plane.

The following notations are equivalent:

$$w(x) = w \begin{pmatrix} x_1 \\ x_2 \end{pmatrix} = \begin{pmatrix} a & b \\ c & d \end{pmatrix} \begin{pmatrix} x_1 \\ x_2 \end{pmatrix} + \begin{pmatrix} e \\ f \end{pmatrix} = \mathbf{A}\,x + \mathbf{t} \quad (4.1)$$

where

$$\mathbf{A} = \begin{pmatrix} a & b \\ c & d \end{pmatrix} \quad (4.2)$$

is a real 2-D matrix and **t** is a vector $\begin{pmatrix} e \\ f \end{pmatrix}$.

The matrix A can be written as:

$$\begin{pmatrix} a & b \\ c & d \end{pmatrix} = \begin{pmatrix} r_1 \cos.\theta_1 & -r_2 \sin.\theta_2 \\ r_1 \sin.\theta_1 & r_2 \cos.\theta_2 \end{pmatrix} \quad (4.3)$$

where r_1, θ_1 are the polar coordinates of point (a, c) and $r_2, (\theta_2 + \pi/2)$ are those of point (b, d).

The affine transform $w(x) = \mathbf{A} x + \mathbf{t}$ in $\mathcal{R}^2$ therefore consists in a linear transform **A**, which deforms the space relative to its origin (rotation then contraction), and a translation of the vector **t** (Figure 4.1).

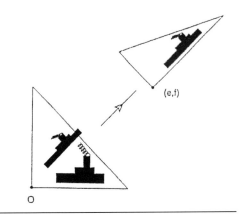

Figure 4.1. Affine transform. This is a linear transform, followed by a translation. From Barnsley, 1988.

Once this transform has been defined, it is possible to look at the computation and drawing of fractal objects using Iterated Function Systems (IFS). Barnsley (1988) proposes two algorithms, determinist and random.

Determinist algorithm from Barnsley (1988)

This algorithm is based on the direct computation of a series of sets:

$$\{\mathcal{A}_n = \mathcal{W}^n (\mathcal{A}_0)\} \quad (4.4)$$

Starting from the initial set $\mathcal{A}_0$, the transform $\mathcal{W}$ allows us to go from $\mathcal{A}_0$ to $\mathcal{A}_1$, then $\mathcal{A}_2$, etc. (Figure 4.2).

One example is the hyperbolic IFS of the shape $\left\{ \mathcal{R}^2;\ w_n:\ n = 1;\ 2;\ ...;\ N \right\}$, where each map is an affine transform w_i.

$$w_i(x) = w_i \begin{bmatrix} x_1 \\ x_2 \end{bmatrix} = \begin{bmatrix} a_i & b_i \\ c_i & d_i \end{bmatrix} \begin{bmatrix} x_1 \\ x_2 \end{bmatrix} + \begin{bmatrix} e_i \\ f_i \end{bmatrix} = A_i + t_i \qquad (4.5)$$

At each iteration:

$$\mathcal{A}_{n+1} = \bigcup_{i=1}^{N} w_i(\mathcal{A}_n) \qquad (4.6)$$

A probability p_i can be associated with w_i. In the general case of an IFS expressed as $\left\{ \mathcal{R}^2;\ w_n:\ n = 1;\ 2;\ ...;\ N \right\}$, there are N numbers $\{p_i:\ i = 1;\ 2;\ ...;\ N\}$ such that $p_1 + p_2 + p_3 + ... + p_N = 1$ and $p_i > 0$ for $i = 1, 2, ..., N$.

These probabilities play an important role in the random algorithm, but do not intervene in the determinist algorithm (if $i = 1, 2, 3$, $p_1 = p_2 = p_3 = 0.33$; see the next codes).

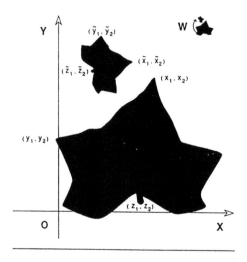

Figure 4.2. Example of a series of affine transforms. The object transformed is an ivy leaf represented in the Euclidean plane (Barnsley, 1988).

Ch. 4] Fractal Sets 163

In the random algorithm, p_i is computed with the determinants of the transform matrices:

$$p_i \approx \frac{|\det \mathcal{A}_i|}{\sum_{i=1}^{N}|\det \mathcal{A}_i|} = \frac{|a_i d_i - b_i c_i|}{\sum_{i=1}^{N}|a_i d_i - b_i c_i|} \qquad (4.7)$$

for i = 1, 2, ..., N.

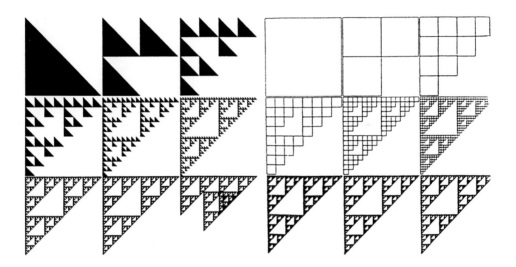

Figure 4.3. Sierpinski's Triangle. IFS results on a triangle and a square, the same determinist algorithm producing the same final figure (Barnsley, 1988).

IFS code for Sierpinski's Triangle

The first example of an IFS code corresponds to Sierpinski's Triangle, a well-known fractal object. The parameters of the code are given in Table 4.1 and its illustration is shown in Figure 4.3.

Table 4.1. IFS code for Sierpinski's Triangle.

w	a	b	c	d	e	f	p
1	0.5	0	0	0.5	1	1	0.33
2	0.5	0	0	0.5	1	50	0.33
3	0.5	0	0	0.5	50	50	0.34

IFS code for Sierpinski's Square

The IFS code for Sierpinski's Square (Table 4.2) starts from a different generator $\mathcal{A}_0$, but produces the same final figure as Sierpinski's Triangle (Figure 4.3).

These first two codes were made for the determinist algorithm (the probability p is the same for each transform w_i).

Table 4.2. IFS code for Sierpinski's Square.

w	a	b	c	d	e	f	p
1	0.5	0	0	0.5	1	1	0.25
2	0.5	0	0	0.5	50	1	0.25
3	0.5	0	0	0.5	1	50	0.25
4	0.5	0	0	0.5	50	50	0.25

IFS Code for a fern leaf

This code, and the next one, will use the random algorithm (Table 4.3).

Table 4.3. IFS code for the fern leaf.

w	a	b	c	d	e	f	p
1	0	0	0	0.16	0	0	0.01
2	0.85	0.04	-0.04	0.85	0	1.6	0.85
3	0.2	-0.26	0.23	0.22	0	1.6	0.07
4	-0.15	0.28	0.26	0.24	0	0.44	0.07

Figure 4.4. Example of application of the IFS code for the fern leaf, with an increasing number of iterations (Barnsley, 1988).

IFS code for a fractal tree

This code is illustrated by Table 4.4 and Figure 4.5.

Table 4.4. IFS code for the fractal tree.

w	a	b	c	d	e	f	p
1	0	0	0	0.5	0	0	0.05
2	0.42	-0.42	0.42	0.42	0	0.2	0.4
3	0.42	0.42	-0.42	0.42	0	0.2	0.4
4	0.1	0	0	0.1	0	0.2	0.15

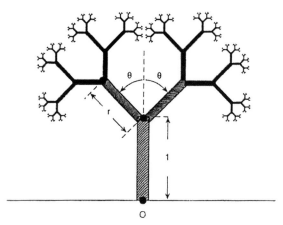

Figure 4.5. Fractal tree, constructed with the IFS code from Table 4.4 (Barnsley, 1988).

4.1.3 Julia Set - Mandelbrot Set

These important examples have been much used to illustrate the result of iterations from the very simple maps which define them.

Complex-based functions as simple as $f(z) = z^2 + c$, where c is a constant, can produce fractals with very original shapes. The Julia sets are obtained by iterations of functions of a complex variable.

General theory

$f: C \to C$ is a polynomial of degree ≥ 2 with complex coefficients:

$$f(z) = a_0 + a_1 z + ... + a_n z^n \qquad (4.8)$$

The successive iterations on f, f o f ...o f, are noted f^k. $f^k(\omega)$ is the k-th iterate $(f(f(...(f(\omega)))))$ of ω.

ω is called a *fixed point* of f if $f(\omega) = \omega$. If $f^p(\omega) = \omega$, p being any integer ≥ 1, ω is called a *periodic point* of f. The smallest p which verifies $f^p(\omega) = \omega$ is called the *period* of ω.

$\{\omega; f(\omega); ...; f^p(\omega)\}$ is called an *orbit* of period p.

Let ω be a periodic point of period p, and $\lambda = (f^p)'(\omega)$, where the symbol (') marks the complex differentiation. ω is called:

$$\begin{cases} \text{over-attracting if } \lambda = 0 \\ \text{attracting if } 0 < \lambda < 1 \\ \text{indifferent if } \lambda = 1 \\ \text{repulsive if } \lambda > 1 \end{cases} \quad (4.9)$$

The Julia set $\mathcal{J}(f)$ can be defined as the closed set composed of the periodic, repulsive points of f.

The complementary set is alled the *Fatou set*, or stable set $\mathcal{F}(f)$. Julia's polynomials are fractal. It can be demonstrated that $\mathcal{J}(f)$ is invariant with f, in both ways: $\mathcal{J} = \mathcal{J}(f) = \mathcal{J}^1(f)$ and $\mathcal{J}$ is a *non-empty compact set*.

f behaves chaotically on $\mathcal{J}$, and $\mathcal{J}$ is generally fractal.

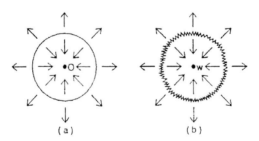

Figure 4.6. Example of Julia sets: (a) for $f(z) = z^2$, the Julia Set is the circle $|z| = 1$, and $f^k(z) \to 0$ when z is inside $\mathcal{J}$; (b) $f(z) = z^2+c$, c small, and $\mathcal{J}$ is fractal (Falconer, 1990).

The simplest example is $f(z) = z^2$, and $f^k(z) = z^{2k}$.

The representative points $z = f^p(z)$ are:

$$\{ \exp(2\pi i q/(2^p-1)) : 0 \le q \le 2^p - 2 \} \tag{4.10}$$

These points are repulsive, as $|(f^p)'(z)| = 2$.

In this case, the Julia set $\mathcal{J}(f)$ is the unit circle $|z| = 1$.

We also have: $\mathcal{J} = \mathcal{J}(f) = \mathcal{J}^1(f)$, with $f^k(z) \to 0$ when $|z| > 1$, but $f^k(z)$ remains on $\mathcal{J}$ for all k if $|z| < 1$.

The Julia set is therefore the boundary between the sets of points that tend, by iteration, toward 0, and those which tend toward ∞. In this particular example (Figure 4.6a), $\mathcal{J}$ is not fractal.

This example can be slightly modified, and the function $f(z)$ replaced by $f(z) = z^2+c$, c being a very small complex number. It is easily observed that $f^k(z) \to \omega$ when z is small, ω being the fixed point of f close to 0, and that $f^k(z) \to \infty$ when z is large.

Once again, the Julia set corresponds to the boundary between two types of behaviour, but, this time, $\mathcal{J}$ is fractal (Figure 4.6b). A series of rigorous demonstrations is given in (Falconer, 1990) and leads to the following conclusions.

The Julia set $\mathcal{J}(f)$ is the closed set composed of the periodic, repulsive points of the polynomial function f. This is a compact uncountable set, containing non-isolated points, and it is invariant on f and f^{-1}. If $z \in \mathcal{J}(f)$, then $\mathcal{J}(f)$ is the closed set formed with $\bigcup_{i=1}^{\infty} f^{-1}(z)$. The Julia set is the limit of the attraction basin of each attracting point of f, ∞ included, and $\mathcal{J}(f) = \mathcal{J}(f^p)$, p positive integer.

Quadratic functions - the Mandelbrot set

Mandelbrot demonstrates how a Julia set of any quadratic function f is the image by h^{-1}, h being a simple linear function, of the Julia set of the polynomial function f_c.

Let $f_c(z) = z^2 + c$ be this polynomial function, $h(z) = \alpha z + \beta$, and h^{-1} its inverse function:

$$h^{-1}(f_c(h(z))) = (\alpha^2 z^2 + 2\alpha\beta z + c - \beta) / \alpha \tag{4.11}$$

With an appropriate choice of α, β, and c, this function can be written as a quadratic function. Therefore:

$$h^{-1} \circ f_c \circ h = f \tag{4.12}$$

And also:

$$h_{-1} \circ f_c^k \circ h = f^k \tag{4.13}$$

168   Illustrations   [Ch.4

for any k.

This means that any series of iterations $f^k(z)$ of a point z with f, will be the exact image by h^{-1} of the series of iterations $f_c^k(h(z))$ of the point h(z) by f_c.

The map h transforms the dynamic image of f into f_c; z is a point of period p of f if h(z) is a point of period p of f_c.

Accordingly, the Julia set of f is the image by h^{-1} of the Julia set of f_c. The transform h is called the combination between f and f_c. Any quadratic function f is conjugate of f_c.

Figure 4.7 shows a few examples of Julia sets obtained for different points c of the Mandelbrot set.

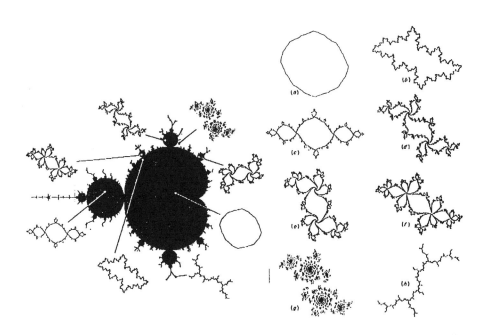

Figure 4.7. Julia sets inside the Mandelbrot set. The sets $\mathcal{J}(f_c)$ are shown for c at different points inside the Mandelbrot set. From Falconer, 1990.

4.2 DYNAMIC SYSTEMS

Two simple constructions illustrate perfectly the important notion of sensitivity to the initial conditions which is characteristic of chaotic systems and attractors.

As the pendulum in a non-central force field, the first example gives a fair idea of chaotic behaviours with an evolution highly dependent on the initial state.

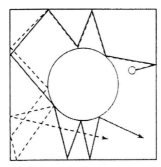

Figure 4.8. Sinaï's billiards. The rapid separation between the two trajectories is an example of strong SIC.

4.2.1 Sinaï's Billiards

The Russian mathematician Sinaï imagined a billiard table in the centre of which is an obstacle of convex shape. In Figure 4.8, the billiard table is square and the obstacle is a circular object laid on the table. Let us observe the trajectories of two balls starting from the same point in conditions apparently identical. Their trajectories will separate rapidly, after a few reflections on the cushion and the circular object, after some reflections, seven or eight, the two trajectories will be completely distinct.

This divergence shows the system's very high sensitivity to the initial conditions. It is intuitively understood that any deviation, however slight and imperceptible, will be very highly amplified.

4.2.2 The Baker Transform

This corresponds to one of the simplest dynamic systems with a fractal attractor. Its name is linked to the similarity of the transform with the repetitive stretching and folding of a piece of dough.

$\varepsilon = [0, 1] \times [0, 1]$ is the square unit. For the fixed value $0 < \lambda < 1/2$, the Baker Transform is defined in ε by:

$$f(x,y) = \begin{cases} (2x, \lambda y) & (0 \le x \le 1/2) \\ (2x - 1, \lambda y + 1/2) & (1/2 \le x \le 1) \end{cases} \quad (4.14)$$

This transform can be said to stretch ε into a rectangle $2 \times \lambda$ by cutting it into two rectangles $1 \times \lambda$, placing then one on top of the other, with a space of $1/2 - \lambda$ (Figure 4.9).

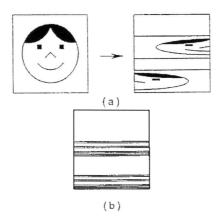

Figure 4.9. The Baker Transform is shown here with (a) the first iteration on the square unit, and (b) its attractor. From Falconer, 1990.

It can easily be demonstrated that this cutting and stretching process leads to an attractor.

If ε_k is the set of order k containing 2^k horizontal strips of width λ^{-k}, spaced by intervals of width $(1/2 - \lambda)^{1-k}$, then: $f(\varepsilon_k) = \varepsilon_{k+1}$.

And the limit set $\mathcal{F} = \bigcap_{k=0}^{\infty} \varepsilon_k$ verifies the relation $f(\mathcal{F}) = \mathcal{F}$.

(x,y) being a point of ε, after k iterations, $f^k(x,y)$ is inside a strip λ^{-k} of $\mathcal{F}$. This means that all points representative of the face (Figure 5.9) are inside the strips. Thus all points of ε are attracted toward $\mathcal{F}$ by the iterations f.

It can also be demonstrated that f is sensitive to the initial conditions, and that $\mathcal{F}$ is fractal, as it is the product $[0, 1] \times \mathcal{F}_1$, where $\mathcal{F}_1$ is a uniform Cantor set obtained by repetitively replacing the intervals I by pairs of sub-intervals of lengths $\lambda|I|$. Using the characteristic property of the product of fractal sets:

$\dim_H. \mathcal{F}_1 = \dfrac{\log.2}{-\log.\lambda}$, and $\dim_H. \mathcal{F} = 1 + \dfrac{\log.2}{-\log.\lambda}$

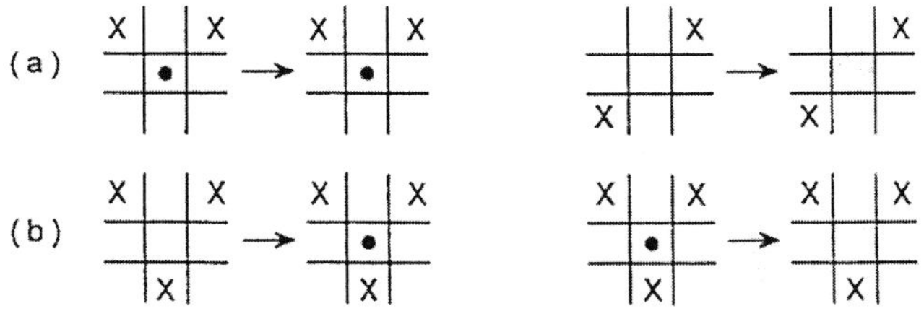

Figure 4.10. The "game of life". The rules of the game (life, birth, death) are explained in the text.

4.3 SELF-ORGANISED CRITICALITY

The examples of the sand heap and dominoes (see Section 3.5) are perfect illustrations of this property characteristic of many natural dynamic systems. The "game of life" is another famous example.

4.3.1 The Game of Life

This game was invented in 1970 by the mathematician J.H. Conway, and simulates the evolution of a colony of living organisms.

To start the game, the organisms are randomly placed on grid formed with square cells. Each cell is occupied by one organism at most, and is surrounded by eight neighbouring cells (Figure 4.10). The status of each site at a particular stage is determined by counting the number of organisms in the eight neighbouring cells.

If there are two living organisms around an empty or occupied cell, the status of this cell does not change.

If there are three living organisms around a given cell, there is birth of a new organism if the cell was empty, or death of an old organism if the cell was occupied.

In any other case, the organism dies by suffocation or isolation.

The game can go on following these rules, until it reaches a stable situation in a simple periodic state where the colonies are stable. When the game is perturbed by the addition of another living cell, the system maintains long periods of transition activity. To quantify these perturbations and their effects, the total activity S is defined as the total number of births and deaths after a given perturbation. Figure 4.11 shows the distribution of the clusterings of average size s over 40,000 perturbations. The distribution $\mathcal{D}(s)$ is a power law: $\mathcal{D}(s) = s^{-t}$, t = 1.4. The distribution $\mathcal{D}(T)$ of the durations T of each perturbation is also a power law: $\mathcal{D}(T) = T^{-b}$, b = 1.6

The fact that activity does not increase or decrease exponentially shows that life and death are highly correlated in time and in space. The system evolved into a critical state and is an example of SOC. It can be remarked that the sites of life are based on a fractal set. The distribution of the number of living sites at a distance r from a given living site increases with r as $\mathcal{D}(r) = r^{D-1}$, r = 1.7.

It was therefore demonstrated that the game of life was functioning in a self-organised critical state, characterised by the critical exponents t, b and D, among others.

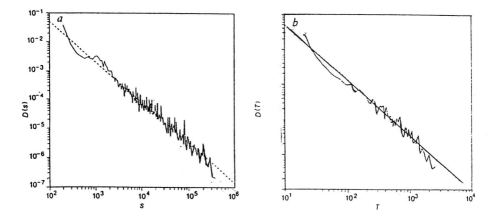

Figure 4.11. In the game of life, log-log representation of the distribution of the size s of clusters and of their duration T. From Bak et al., 1989.

A game is an admirable thing to devise
Paul Valéry, 1925

Annex A - Fractal Theory

> *Borel told me the other day that he had to give up his research on the theory of sets because of the exhaustion it gave him, and it made him fear for more serious trouble if he were to persist. He added that Cantor and Baire had already been seriously sick following their efforts in the field. I do not know of any poet which went to these excesses.*
> Paul Valéry, 1924

A.1 MEASURE THEORY

The notions of measure and dimension are fundamental not only for the study of fractal and multifractal sets, but also for the study of dynamic systems.

A.1.1 Definition of the Hausdorff Measure

This definition was given to me by the mathematician J.P. Kahane (personal communication, 1989).

Let h be an increasing function from $\mathcal{R}^+$ to $\mathcal{R}^+$. Let $\mathcal{E}$ be a metric space, $\mathcal{A}$ a part of $\mathcal{E}$. The Hausdorff measure of $\mathcal{A}$ relative to h is:

$$\mu_h(\mathcal{A}) = \lim_{\delta \to 0} \mu_{h,\delta}(\mathcal{A}) \tag{A.1}$$

where:

$$\mu_{h,\delta}(\mathcal{A}) = \inf. \sum_i h(\text{diameter}(\mathcal{B}_i)) \qquad (A.2)$$

The lower limit is taken for all covers of $\mathcal{A}$ by balls $\mathcal{B}_i$ of diameter $\leq \delta$. It can be noted that: $0 \leq \mu_{h,\delta}(\mathcal{A}) \leq \infty$, and that $\mu_{h,\delta}$ increases when δ decreases.

Thus, $0 \leq \mu_h(\mathcal{A}) \leq \infty$. μ_h is an external measure, i.e. it verifies the properties $\mu_h(\emptyset) = 0$ and $\mu_h(\bigcup_{i=1}^{\infty} \mathcal{A}_n) \leq \sum_{n=1}^{\infty} \mu_h(\mathcal{A}_n)$. The sets measurable with μ_h are the parts $\mathcal{A}$ of $\mathcal{E}$ such that:

$$\mathcal{B} \subset \mathcal{A} \text{ and } C \subset \mathcal{E} \setminus \mathcal{A} \Rightarrow \mu_h(\mathcal{B} \cup C) = \mu_h(\mathcal{B}) + \mu_h(C) \qquad (A.3)$$

These sets make up a "tribe", and μ_h is a measure in the usual sense on this "tribe". It is totally measurable all Borelians of $\mathcal{E}$ are measurable with μ_h.

If $\mathcal{E}$ is $\mathcal{R}^n$, the measure μ_h is the extenal measure of Lebesgue (see Section 1.1, and further sections).

A.1.2 About the Notions of Measure and Dimension

In order to understand correctly the notions of measure, let us examine the definitions of the Lebesgue and Hausdorff measures presented in Section 1.1:

$$\mathcal{L}^n(\mathcal{A}) = \inf. \left\{ \sum_{i=1}^{\infty} \text{vol}^n(\mathcal{A}_i) \, ; \, \mathcal{A} \subset \bigcup_{i=1}^{\infty} \mathcal{A}_i \right\} \qquad (A.4)$$

$$\mathcal{H}^s_\delta(\mathcal{F}) = \inf. \left\{ \sum_{i=1}^{\infty} |\mathcal{U}_i|^s \, ; \, \mathcal{U}_i \text{ being a } \delta\text{-cover of } \mathcal{F} \right\} \qquad (A.5)$$

and its limit when $\delta \to 0$ is the Hausdorff measure dimension of dimension s of $\mathcal{F}$.

$$\mathcal{H}^s(\mathcal{F}) = \lim_{\delta \to 0} \mathcal{H}^s_\delta(\mathcal{F}) \qquad (A.6)$$

Let us now measure on $\mathcal{R}^n$, n integer. To introduce the Hausdorff measure, we shall first look at the notion of *restriction of a measure on $\mathcal{R}^n$*. Let μ be a measure on $\mathcal{R}^n$, and $\mathcal{E}$ a Borelian subset of $\mathcal{R}^n$. For any set $\mathcal{A}$, a measure ν can be defined as the restriction of μ to $\mathcal{E}$ with: $\nu(\mathcal{A}) = \mu(\mathcal{E} \cap \mathcal{A})$. ν is a measure on $\mathcal{R}^n$, whose base is contained in $\overline{\mathcal{E}}$ (closed

Annex A] Fractals 175

$\mathcal{E}$). Such are the Hausdorff measures $\mathcal{H}^s$ of dimension s on subsets of $\mathcal{R}^n$ with $0 \leq s \leq n$. This measure of dimensions, which can be non-integer, is therefore a generalisation of the Lebesgue measure (which is expressed for spaces with integer dimensions) to subsets $\mathcal{A}$ of $\mathcal{R}^n$.

In the case where the subset of $\mathcal{R}^n$ is of integer dimension, the Hausdorff measure is the same as the Lebesgue measure, with a multiplicative factor (see Section 1.1).

A.1.3 About the Hausdorff Dimension

We will develop here the illustration of the notion of non-integer Hausdorff dimension given by Manneville (1991). In a space of dimension d (Euclidean, d integer), the mass of a compact ball of radius ε varies as ε^d. The mass of an ε-cover of a set is given by: $M_d \sim N(\varepsilon)\varepsilon^d$. And, as $N(\varepsilon) \sim \varepsilon^{-d_f}$, where d_f is the fractal dimension of the set, $M_d \sim \varepsilon^{d-d_f}$, for $\varepsilon \to 0$.

If $d_f < d$, $M_d \to 0$ when $\varepsilon \to 0$.

Let us choose a test dimension $d' \leq d$, and cover the set with balls of dimension d' (for example, disks, for which d' = 2, in an Euclidean space where d = 3). Reasoning as previously, we have: $M_{d'} \sim \varepsilon^{d'-d'_f}$.

$$d' > d_f \text{ when } \varepsilon \to 0, M_{d'} \to 0$$

Therefore, when $\varepsilon \to 0$, $d' < d_f$ $M_{d'} \to \infty$

$d' = d_f$ $M_{d'} \to $ constant

The fractal dimension is therefore the dimension d' for which the d'-mass has a finite limit.

A.2 FRACTALS

From the paradise Cantor created for us,
None should chase us.
                                Hilbert, 1930

A.2.1 Hölder Function, Hölder Condition, Lipschitz Map

(1) A function f: X $\to$ Y is called a Hölder function of exponent α if:

$$|f(x) - f(y)| \leq c |x - y|^\alpha, (x,y \in X) \tag{A.7}$$

where c is a constant. The function f is a Lipschitz function if α can be equal to 1, and a Lipschitz bi-function if:

$$c_1 |x - y| \leq |f(x) - f(y)| \leq c_2 |x - y|, \quad (x,y \in X) \tag{A.8}$$

for $0 < c_1 \leq c_2 < \infty$

(2) Let f: $\mathcal{F} \to \mathcal{R}^n$ ($\mathcal{F} \subset \mathcal{R}^n$) be a map such that:

$$|f(x) - f(y)| \leq c |x - y|^\alpha, \quad (x,y \in X) \tag{A.9}$$

where $c > 0$ and $\alpha > 0$ are constants.

Then, for all values of s,

$$\mathcal{H}^{s/\alpha}(f(\mathcal{F})) \leq c^{s/\alpha} \mathcal{H}^s(\mathcal{F}) \tag{A.10}$$

This relation is demonstrated by considering the δ-covers $\{\mathcal{U}_i\}$ of $\mathcal{F}$.

$$|f(\mathcal{F} \cap \mathcal{U}_i)| \leq c |\mathcal{U}_i|^\alpha \tag{A.11}$$

From this, it comes that $\{f(\mathcal{F} \cap \mathcal{U}_i)\}$ is an ε-cover of $f(\mathcal{F})$ with $\varepsilon = c\delta^\alpha$, and:

$$\sum_i |f(\mathcal{F} \cap \mathcal{U}_i)|^{s/\alpha} \leq c^{s/\alpha} \sum_i |\mathcal{U}_i|^s \tag{A.12}$$

and

$$\mathcal{H}^{s/\alpha}_\varepsilon (f(\mathcal{F})) \leq c^{s/\alpha} \mathcal{H}^s_\delta(\mathcal{F}) \tag{A.13}$$

with $\delta \to 0$ and $\varepsilon \to 0$.

(3) Let us suppose that f: $\mathcal{F} \to \mathcal{R}^n$ ($\mathcal{F} \subset \mathcal{R}^n$) verifies the Hölder Condition:

$$|f(x) - f(y)| \leq c |x - y|^\alpha, \quad (x,y \in X) \tag{A.14}$$

then: $\dim_H . f(\mathcal{F}) \leq (1/\alpha) \dim_H . \mathcal{F}$

Indeed, if $s > \dim_H . \mathcal{F}$, we get from the preceding demontration:

$$\mathcal{H}_\varepsilon^{s/\alpha}(f(\mathcal{F})) \leq c^{s/\alpha}\, \mathcal{H}_\delta^s(\mathcal{F}) = 0$$

which implies that $\dim_H.f(\mathcal{F}) \leq s/\alpha$, for any $s > \dim_H.\mathcal{F}$

(4) If $f: \mathcal{F} \to \mathcal{R}^n$ is a Lipschitz transform, then $\dim_H.f(\mathcal{F}) \leq \dim_H.\mathcal{F}$

(5) If $f: \mathcal{F} \to \mathcal{R}^n$ is a Lipschitz bi-transform, then:

$$c_1 |x - y| \leq |f(x) - f(y)| \leq c_2 |x - y|, (x,y \in X) \quad (A.15)$$

where $0 < c_1 \leq c_2 < \infty$, and then: $\dim_H.f(\mathcal{F}) = \dim_H.\mathcal{F}$

A.2.2. Mandelbrot Set on Quaternions

The figure on Colour Plate 1, between pp. 192 and 193, was computed on a Mandelbrot set by Stoclet (personal communication) using an algorithm written by Norton (1982).

The body of quaternions is defined by:

$$z = z_1 + z_2\, i + z_3\, j + z_4\, k \quad (A.16)$$

where $i^2 = j^2 = k^2 = 1^2 = -1$, and $ij = -ji = k$, $jk = -kj = i$, $ki = -ik = j$.

The addition is the one defined on $\mathcal{R}^4$.

The multiplication is not commutative, and:

$$\begin{aligned}
zu \times zv = &\; zu_1 zv_1 - zu_2 zv_2 - zu_3 zv_3 - zu_4 zv_4 \\
& + (zu_1 zv_2 + zu_2 zv_1 + zu_3 zv_4 - zu_4 zv_3)\, i \\
& + (zu_1 zv_3 + zu_3 zv_1 + zu_4 zv_2 - zu_2 zv_4)\, j \\
& + (zu_1 zv_4 + zu_4 zv_1 + zu_2 zv_3 - zu_3 zv_2)\, k
\end{aligned} \quad (A.17)$$

if $zu = zu_1 + zu_2\, i + zu_3\, j + zu_4\, k$ and $zv = zv_1 + zv_2\, j + zv_3\, j + zv_4\, k$.

The conjugate of z is $z^* = z_1 - z_2\, i - z_3\, j - z_4 k$.

The module of z is: $|z|^2 = z \times z^*$, i.e. $|z|^2 = z_1^2 + z_2^2 + z_3^2 + z_4^2$.

The escaping times for the Mandelbrot set are $z^{(n+1)} = (z^{(n)})^2 + c$, $z^{(0)} = c$ for the quaternions, with the restriction $z_4 = 0$, and starting from the regular 3-D grid c, of origin (x_0, y_0, z_0):

$$cx(i, j, k) = x_0 + (i - 1) * hx$$
$$cy(i, j, k) = y_0 + (i - 1) * hy \quad (A.18)$$
$$cz(i, j, k) = z_0 + (i - 1) * hz$$

We get:

$$z_1^{(n+1)}(i, j, k) = z_1^{(n)}(i, j, k)^2 - z_2^{(n)}(i, j, k)^2 - z_3^{(n)}(i, j, k)^2 + cx(i, j, k)$$
$$z_2^{(n+1)}(i, j, k) = 2z_1^{(n)}(i, j, k) z_2^{(n)}(i, j, k) + cy(i, j, k) \quad (A.19)$$
$$z_3^{(n+1)}(i, j, k) = 2z_1^{(n)}(i, j, k) z_3^{(n)}(i, j, k) + cz(i, j, k)$$

A.2.3 Fractal Projections

Projection in $\mathcal{R}^2$

Using the preceding developments, the theorem of Section 1.3.8 can be easily demonstrated by remarking that:

$$|Pr_\theta x - Pr_\theta y| \leq |x - y| \quad (A.20)$$

This relation shows that the function Pr_θ is a Lipschitz map. And we saw earlier that, if such is the case for function f, then:

$$\dim_H . f(\mathcal{F}) \leq \dim_H . \mathcal{F} \quad (A.21)$$

Therefore:

$$\dim_H . (Pr_\theta \mathcal{F}) \leq \dim_H . \mathcal{F} \quad (A.22)$$

And, as this dimension is at most equal to 1, we finally get:

$$\dim_H . (Pr_\theta \mathcal{F}) \leq \min . \left\{ \dim_H \mathcal{F}; 1 \right\} \quad (A.23)$$

Local density; regular and irregular sets; geometrical implications

The notions of regular and irregular sets are important for the study of projections, products and intersections of fractal sets. Let us first define the density around particular points of a fractal set.

Density

If $\mathcal{F}$ is a subset of the plane, the density of $\mathcal{F}$ at point x is given by:

$$\lim_{r \to 0} \frac{\text{area}(\mathcal{F} \cap \mathcal{B}_r(x))}{\text{area}(\mathcal{B}_r(x))} = \lim_{r \to 0} \frac{\text{area}(\mathcal{F} \cap \mathcal{B}_r(x))}{\pi r^2} \qquad (A.24)$$

where $\mathcal{B}_r(x)$ is a disk of radius r and centre x.

Lebesgue's density theorem says that, for a Borelian set, this limit exists and equals 1 when $x \in \mathcal{F}$, and equals 0 when $x \notin \mathcal{F}$, provided that the set $\{x\}$ does not have a null area.

In the case where $\mathcal{F}$ is a continuous curve in the plane, and x is a point of this curve (different from its extremities), $\mathcal{F} \cap \mathcal{B}_r(x)$ is close to a string of $\mathcal{F}$ and:

$$\lim_{r \to 0} \frac{\text{length}(\mathcal{F} \cap \mathcal{B}_r(x))}{2r} = 1 \qquad (A.25)$$

If $x \notin \mathcal{F}$, this limit is 0.

It is possible to study sets of dimension s, and investigate the Hausdorff measure of dimension s when $r \to 0$.

Let $\mathcal{F}$ be a s-set of $\mathcal{R}^2$ where $0 < s < 2$ (the limit cases are a set of points, for $s = 0$, and the preceding case, for $s = 2$).

The lower and upper densities of a s-set $\mathcal{F}$ at a point $x \in \mathcal{R}^n$ are defined by:

$$\underline{D}^s(\mathcal{F},x) = \lim_{r \to 0} \frac{\mathcal{H}^s(\mathcal{F} \cap \mathcal{B}_r(x))}{(2r)^s} \qquad (A.26)$$

and:

$$\overline{D}^s(\mathcal{F},x) = \overline{\lim_{r \to 0}} \frac{\mathcal{H}^s(\mathcal{F} \cap \mathcal{B}_r(x))}{(2r)^s} \qquad (A.27)$$

If $\underline{D}^s = \overline{D}^s$, the density of $\mathcal{F}$ exists and is written as: $D^s(\mathcal{F},x)$.

Regular and irregular sets

A point $x \in \mathcal{F}$ is called a *regular point* of $\mathcal{F}$ if it verifies: $\underline{D}^s(\mathcal{F},x) = \overline{D}^s(\mathcal{F},x) = 1$.

If it does not, the point x is called an *irregular point*.

A s-set is said to be regular if "$\mathcal{H}^s$ - nearly all its points" are regular.

A s-set is irregular if "$\mathcal{H}^s$ - nearly all its points" are irregular.

A fundamental result is that an s-set of $\mathcal{F}$ is necessarily irregular, except if s is an integer.

Still using this notion of density, it is possible to demonstrate that, if $\mathcal{F}$ is a s-set of $\mathcal{R}^2$ (Falconer, 1990):

$$\underline{D}^s(\mathcal{F},x) = \overline{D}^s(\mathcal{F},x) = 0 \tag{A.28}$$

for "$\mathcal{H}^s$ - nearly all points $x \notin \mathcal{F}$", and:

$$2^{-s} \leq \overline{D}^s(\mathcal{F},x) \leq 1 \tag{A.29}$$

for "$\mathcal{H}^s$ - nearly all points $x \in \mathcal{F}$".

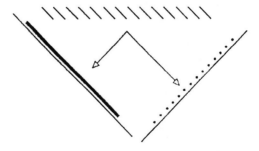

Figure A.1. Example of the Venetian blind. Depending on the direction of projection, the length can be null or of dimension 1 (Falconer, 1990).

An interesting case is the structure of 1-sets, with an integer dimension but a complex structure which can be decomposed into a regular and an irregular part. Figure A.1 shows an example, where $\mathcal{F}$ can be decomposed into a curve or set of curves, and points for the irregular part. The projection of a set of this type leads to three possibilities:
- If $\mathcal{F}$ is a regular "1-set" of $\mathcal{R}^2$, then the projection $\text{Pr}_\theta \mathcal{F}$ on the line of angle θ has a positive length (i.e. measure), for any $\theta \in [0; \pi]$ (except for at most one value of θ).
- $\mathcal{F}$ is a "1-set" of $\mathcal{R}^2$. If its regular part has an $\mathcal{H}^1$-measure of 0, then $\text{Pr}_\theta \mathcal{F}$ has a null length for nearly all θ. Otherwise, it has a positive length for all values of θ, except at

most one.
- A "1-set" is irregular if, and only if, its projections have null lengths in at least two directions.

The example of the Venetian blind in Figure A.1 corresponds to the second case.

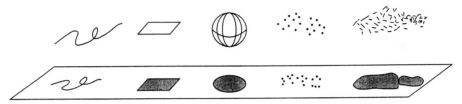

Projection plane

Figure A.2. Projection on a plane of various geometric objects and fractal sets defined in $\mathcal{R}^3$. The fractal set $\mathcal{F}_1$ has a dimension $\dim_H. \mathcal{F}_1 < 1$ and its projection onto the plane has a dimension of $\dim. \mathcal{F}_1$. The fractal set $\mathcal{F}_2$ has a dimension $\dim_H. \mathcal{F}_2 > 1$ and its projection has a dimension of 2.

Another example, shown in Figure A.2, is an extension to $\mathcal{R}^3$. Different geometric objects with integer dimensions are projected onto a plane. The curve of dimension 1 produces on the plane a curve of dimension 1 and a positive length (measure). A surface and a cube have projections of dimension 2 and positive areas (measures). The fractal sets $\mathcal{F}_1$ and $\mathcal{F}_2$, of respective dimensions $\dim_H. \mathcal{F}_1 < 1$ and $\dim_H. \mathcal{F}_2 > 1$, the projections onto the plane respectively have a dimension $\dim_H. \mathcal{F}_1$ and a null area, and a dimension of 2 and a positive area.

A.3 DYNAMIC SYSTEMS AND CHAOS

Time: duration, continued existence;
progress of this viewed as
affecting persons or things
Concise Oxford Dictionary

A.3.1 Degree of Freedom, Phase Space, Phase Trajectories

These terms have been used all through Section 1.4. For their rigorous definition, the reader is refered to the classic books of theoretical physics written some time ago by Landau and Lifschitz (1967, 1969).

Degree of freedom

To determine the position of a system with N points in space, one needs N radial vectors, i.e. 3N coordinates for a 3-D space. The number of independent values needed to determine the system's position without ambiguity is called the system's *number of degrees of freedom*.

These values are not necessarily Cartesian coordinates of a point. s values $q_1, q_2, ..., q_s$ which define completely the position of a system with s degrees of freedom are called *generalised coordinates*, and their derivatives q_i' the *generalised velocities*. Experiment shows that the simultaneous knowledge of coordinates and velocities enables the full determination of the system's state, and, in principle, allows us to predict its future behaviour.

Phase space - phase trajectories

We will take the definitions of Landau and Lifschitz (1967) again. Let us consider a macroscopic mechanical system with s degrees of freedom (and these definitions can extend to any dynamic system). The position of each point in space (physical space) is characterised by s coordinates q_i, i = 1, ..., s. At a given time, the state of the system can be entirely determined by the simultaneous values of the s coordinates q_i and of the corresponding velocities q_i'. In statistics, the pulses p_i are used instead of velocities. Each system thus has a specific phase space with 2n dimensions, where n is the number of degrees of freedom.

Any point in the phase space corresponding to given values of the system's coordinates q_i and impulses p_i represents a determined state of the system. When the system's state varies with time, the corresponding point in the phase space will follow a curve called the *phase trajectory*.

A.3.2 Chaos Applied to Dynamic Systems

Let us consider a field of vectors C^r ($r \geq 1$) and maps of $\mathcal{R}^n$ defined by:

$$\text{the field of vectors} \quad x' = f(x) \tag{A.30}$$

$$\text{the map} \quad x \mapsto g(x) \tag{A.31}$$

The corresponding flow is $\Phi(t,x)$, and is assumed to exist for any $t > 0$. $\Lambda \subset \mathcal{R}^n$ is a compact set invariant under $\Phi(t,x)$ (respectively g(x)). $\Phi(t,\Lambda) \subset \Lambda$ for any $t \in \mathcal{R}$ (respectively $g^n(\Lambda) \subset \Lambda$ for any $n \in \mathcal{Z}$, except if g cannot be inverted, in which case n must be ≥ 0). Then we can set the two following definitions.

First definition

The flow $\Phi(t,x)$ is said likely to be sensitive to the initial conditions on Λ, if there exists $\varepsilon \geq 0$, such that, for any $x \in \Lambda$, and for any neighbourhood U of x, there is a $y \in U$ and t > 0, which verify: $|\Phi(t,x) - \Phi(t,x)| > \varepsilon$.

Respectively, g(x) is sensitive to the initial conditions on Λ, if there exists $\varepsilon \geq 0$, such that, for any $x \in \Lambda$, and for any neighbourhood U of x, there exists $y \in U$ and $n \geq 0$, which verify: $|g^n(x) - g^n(y)| > \varepsilon$.

This means that in any point $x \in \Lambda$, there is at least one point arbitrarily close to Λ which diverges from x. The rate of divergence is considered by some authors to be exponential, but this is not necessary.

Second definition

Λ is chaotic if:

- $\Phi(t,x)$ (respectively g(x)) is sensitive to the initial conditions on Λ.

- $\Phi(t,x)$ (respectively g(x)) is topologically transitive on Λ.

A third condition is sometimes added:

- The periodic orbits of $\Phi(t,x)$ (respectively g(x)) are dense in Λ.

A.3.3 Control of Chaotic Systems

Chaotic systems are characterised by a very high sensitivity to the smallest perturbations. This property was defined in Section 1.7 and in Section A.3 under the name of SIC, Sensitivity to the Initial Conditions. It is also known as the "butterfly effect", a concept much presented in the media.

This sensitivity makes it possible to control chaotic systems in a way not attainable with stable systems. In the latter, small interventions only induce small changes. In a chaotic system, however, it is possible to choose among many dynamic behaviours when one knows the system well.

Shinbrot et al. (1993) suggest the stabilising of unstable orbits. Indeed, one of the major characteristics of chaos is that several possible movements are present together in the system. Some of the orbits are periodic, and if the system would be stable if it were moving to these orbits. But these orbits are themselves unstable, as the slightest deviation from them (for example because of noise) increases exponentially with time, and the system rapidly moves away from the periodic orbit. This notion is better understood with the simple example given by Shinbrot et al. (1993).

They choose Hénon's map, defined as a 2-D map where the state of the system at instant n (n = 0, 1, 2 ...) is determined by two scalar variables x_n and y_n. The map specifies the way to go from one state of the system, at instant n, to the state at time n+1. We saw in Section 1.5.5 that Hénon's map is defined by:

$$x_{n+1} = p + 0.3\, y_n - x_n^2 \qquad (A.32)$$

$$y_{n+1} = x_n$$

where p is the control parameter, and is given the value $p_0 = 1.4$.

From the starting point (x_1, y_1), we can compute (x_2, y_2), (x_3, y_3), etc.

These iterations converge onto a strange attractor (see Section 1.5.5), with a fractal dimension of 1.26. Figure A.3 shows some of the sections with unstable periodic orbits. The point A_* of period 1 is revisited at each iteration; the points B_1 and B_2 of period 2 are revisited alternatively ($B_1 \to B_2 \to B_1 \to B_2$); and the points C_1, C_2, C_3 and C_4 of period 4 are revisited every 4 iterations.

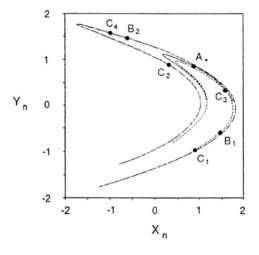

Figure A.3. Poincaré section of Hénon's attractor. A_* is the point of period 1; B_1 and B_2 are the points of period 2; C_1, C_2, C_3 and C_4 are the points of period 4. From Shinbrot et al., 1993.

This is a map, and time therefore intervenes as a discrete value. A dynamic system generally has systems of M first-order differential equations, of the type: $d\xi/dt = G(\xi)$, where $\xi(t) = (\xi^{(1)}(t), \xi^{(2)}(t), ..., \xi^{(M)}(t))$ is a vector and time is a continuous variable.

Discrete-time systems can also be interesting, as the continuous-time system with M dimensions can be reduced to (M-1) dimensions through Poincaré sections, as shown in Figure A.4 for M = 3. Figure A.4 represents a continuous-time trajectory and its discrete-time equivalent Z_1, Z_2, ... Z_n represents the vector with M-1 coordinates (2 in this case), which defines the position on the section's surface of the n-th crossing point of the upward trajectory. For a given value of Z_n, the equation $d\xi/dt = G(\xi)$ can be integrated before this point and up to the next point crossing the surface, Z_{n+1}. Z_{n+1} is uniquely determined by Z_n, and there must be a map $Z_{n+1} = F(Z_n)$ which goes from one point of the trajectory on the section, to the next.

Even if **F** cannot be explicitly determined, its very existence is useful. Figure A.4 shows an orbit of period 1 for a map of dimension N, $Z_{n+1} = F(Z_n, p)$, where **Z** is an N-dimensional vector and p a control parameter. For values of p close to p_0, and close to the point of period 1, $Z_* = F(Z_*, p_0)$, the dynamics is approximated with a linear map:

$$(Z_{n+1} - Z_*) = A.(Z_n - Z_*) + B\,(p - p_0) \tag{A.33}$$

where **A** is the NxN Jacobian matrix (see Section 1.5.6) and **B** is a vector of dimension N, where $A = \partial F/\partial Z$, $B = \partial F/\partial p$ and these partial derivatives are evaluated for $Z = Z_*$ and $p = p_0$.

Let us now suppose that p can be adjusted at each iteration. Z_n is determined, and p is slightly changed from the nominal value p_0. p is replaced by p_n. Considering that the relation is linear, we get:

$$(p_n - p_0) = -K^T.(Z_n - Z_*) \tag{A.34}$$

where **K** is a constant vector with N dimensions and K^T its transposed vector. The choice of **K** implies a specification of p_n for each iteration.

Substituting equation (A.34) in equation (A.33), we get:

$$\delta Z_{n+1} = (A - B.K^T)\,\delta Z_n \tag{A.35}$$

where $\delta Z_n = Z_n - Z_*$.

Thus the point Z_* of period 1 will be stable if **K** can be chosen so that the eigenvalues of matrix $(A - B.K^T)$ have modules smaller than 1. Then $\delta Z_n \to 0$, i.e. $Z_n \to Z_*$ when n tends toward ∞. The choice of **K** is the typical problem of chaos control theory.

This procedure gives the values of the time-dependent parameter p_n, necessary to stabilise an unstable point of period 1 in a chaotic system. Because of practical constraints, the deviations of p_n relative to p_0 should not be too large; $|p_n - p_0|$ is limited by a maximum value $\delta p_{max} > |K^T.(Z_n - Z_*)|$. If the system's state falls outside this region, no perturbation is applied and one waits until the system's state falls back inside this region, to apply the slight perturbation satisfying equation (A.34).

As an example, Shinbrot et al. (1993) show the stabilisation of a periodic orbit going through A_*, on Hénon's attractor, by adjustments of p at 1% of its nominal value. The starting point has been randomly chosen on the attractor. It can be seen on Figure A.5 that, for the first 86 iterations, the trajectories are chaotically moving on the attractor without falling in the small region of A_* which needs to be reached. At the 87th iteration, the system's state falls exactly into the region desired, and is then maintained close to A_*.

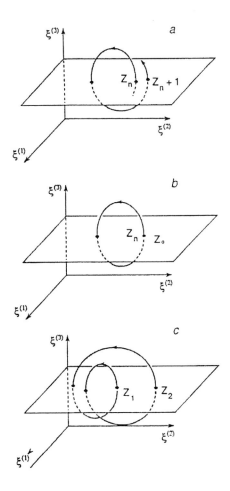

Figure A.4. Poincaré sections of a few trajectories, with ξ_3 constant. The trajectories correspond to orbits of period 1 in (b), of period 2 in (c) and of any period in (a) (Shinbrot et al., 1993).

Figure A.6 illustrates how a periodic orbit, the matrix **A** and the vector **B** can be extracted from the observation of the trajectory on the attractor. $\mathbf{Z}_1$, $\mathbf{Z}_2$, etc. are the points of intersection with the surface. If two successive $\mathbf{Z}_s$ are close to each other (for example $\mathbf{Z}_{100}$ and $\mathbf{Z}_{101}$), there must be an orbit $\mathbf{Z}_*$ of period 1 at proximity (Figure A.6). After the observation of a first close return, one looks for the next pairs of close returns ($\mathbf{Z}_n$, $\mathbf{Z}_{n+1}$) in this region. If the data series is long enough, there will be many identical pairs.

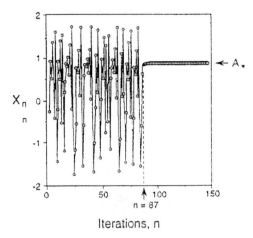

Figure A.5. Mastering chaos. Example of the stabilisation of a system in a state of period 1 by using a Hénon map (Shinbrot et al., 1993).

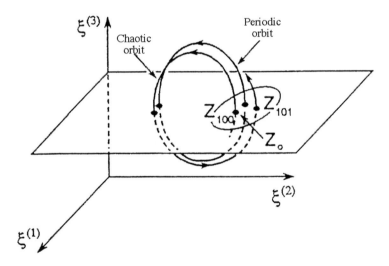

Figure A.6. Looking for the stable orbit of period 1 is equivalent to looking for a point $\mathbf{Z}_*$ with the data $\mathbf{Z}_{100}$, $\mathbf{Z}_{101}$, ... (Shinbrot et al., 1993)

Assuming the region is small, the pairs of close returns can be adjusted with a linear relation:

$$\mathbf{Z}_{n+1} = \underline{\mathbf{A}}.\mathbf{Z}_n + \underline{\mathbf{C}} \qquad (A.36)$$

Generally, if the data is noisy, the largest possible number of pairs will be used to adjust the matrix $\underline{\mathbf{A}}$ and the vector $\underline{\mathbf{C}}$ by a least-squares regression. The least-squares adjustement matrix $\underline{\mathbf{A}}$ is an approximation of the Jacobian matrix $\mathbf{A}$ of equation (A.33). And the point $\mathbf{Z}_*(p)$ of period 1 is approximately given by $(1-\underline{\mathbf{A}})^{-1}\underline{\mathbf{C}}$. To find the vector $\mathbf{B}$ of equation (A.33), p is slightly modified: $p \mapsto p+\Delta p$, and the point of period 1 $\mathbf{Z}_*(p+\Delta p)$ is determined again. $\mathbf{B}$ is approximated with $[\mathbf{Z}_*(p+\Delta p) - \mathbf{Z}_*(p)] / \Delta p$.

The orbits of period 2 are found in the same way, but looking at pairs $(\mathbf{Z}_n, \mathbf{Z}_{n+2})$, and so on for orbits of period 4, etc.

A.4 MULTIFRACTALS

By set, we mean a grouping of objects distinct in our intuition or our thought
Cantor, 1890

During the determination of singularity spectra for multifractal sets, the methods are generally tested on models whose fractal base is a Cantor set (see Section 1.6). This allows to use cascades up to the value of k necessary to determine the exponent α between α and $\alpha+d\alpha$. In reality, there is no possibility to use multiplicative cascades. Methods must therefore be established for the empirical estimation of $f(\alpha)$ for a given measure. With Evertsz and Mandelbrot (1990), we propose here two practical methods which can be added to the ones exposed in Section 2.6 (Geilikman et al., 1990; Jensen et al., 1985).

A.4.1 The Histogram Method

For a measure μ, the histogram method can be separated into the following stages:

1. Distribute the measure coarsely with boxes of size ε. This leads to a series of boxes:

$$\{B_i(\varepsilon)\}_{i=1}^{N(\varepsilon)} \tag{A.37}$$

where $N(\varepsilon)$ is the number of boxes needed to cover the set supporting the measure.

2. If $\mu(B_i)$ is the measure of box i, we can compute the raw Hölder exponent:

$$\alpha_i = \log.\mu_i / \log.\varepsilon \tag{A.38}$$

3. Then, we build the histogram, in which the variable α is divided into segments of size $\Delta\alpha$ suitably chosen. And we estimate $N_\varepsilon(\alpha)$ by counting the number of times $N_\varepsilon(\alpha)\Delta\alpha$ that a specific value falls between α and $\alpha+d\alpha$.

4. The process is repeated for different values of the size ε.

5. We are looking for a relation similar to:

$$N_\varepsilon(\alpha) \sim \varepsilon^{-f(\alpha)} \tag{A.39}$$

This is made by plotting $-\log.N_\varepsilon(\alpha) / \log.\varepsilon$ as a function of α for several values of ε.

If these points fall on a curve $f(\alpha)$, for small enough values of ε, *the measure is multifractal*.

A.4.2 The Moments Method

This method is based on a function called *allotment function*, which is defined as:

$$\chi_q(\varepsilon) = \sum_{i=1}^{N(\varepsilon)} \mu_i^q \; , q \in \mathcal{R} \tag{A.40}$$

In the case of the binomial measure (analysed in Section 1.6), the allotment function is $\chi_q(\varepsilon_k)$, corresponding to the box size $\varepsilon_k = 2^{-k}$. We have:

$$\chi_q(\varepsilon_0) = 1^q,$$

$$\chi_q(\varepsilon_1) = m_0^q + m_1^q,$$

$$\chi_q(\varepsilon_2) = (m_0^q + m_1^q)^2,$$

and more generally $\chi_q(\varepsilon_k) = (m_0^q + m_1^q)^k$ (cf. Figure A.7).

In the previous definition, the measures μ_i of boxes can be written as $\mu_i = \varepsilon^{\alpha_i}$. Therefore:

$$\chi_q(\varepsilon) = \sum_{i=1}^{N(\varepsilon)} (\varepsilon^{\alpha_i})^q \tag{A.41}$$

Let $N_\varepsilon(\alpha)d\alpha$ be the number of boxes, out of the total number $N(\varepsilon)$, for which the raw Hölder exponent satisfies: $\alpha < \alpha_i < \alpha + d\alpha$. Let us assume there are constants α_{min} and α_{max} such that: $0 < \alpha_{min} < \alpha < \alpha_{max} < \infty$, and that $N_\varepsilon(\alpha)$ is continuous. Then the contribution to $\chi_q(\varepsilon)$ of the subset of boxes with $\alpha_i \in [\alpha; \alpha+d\alpha]$ is $N_\varepsilon(\alpha)(\varepsilon^\alpha)^q \, d\alpha$. We can integrate on $d\alpha$ the contributions of all these subsets whose raw Hölder exponent is between α and $\alpha+d\alpha$. For a set value of ε:

$$\chi_q(\varepsilon) = \int N_\varepsilon(\alpha) \, (\varepsilon^\alpha)^q \, d\alpha \tag{A.42}$$

If $N_\varepsilon(\alpha) \sim \varepsilon^{-f(\alpha)}$, we get:

$$\chi_q(\varepsilon) = \int \varepsilon^{q\alpha - f(\alpha)} \, d\alpha \tag{A.43}$$

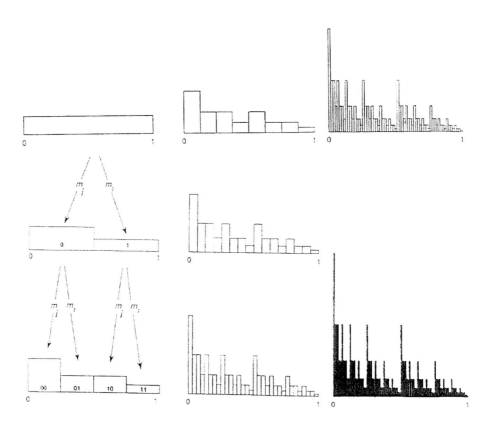

Figure A.7. Example of a cascade generating a binomial measure. We chose here a dyadic set where the measure verifies $m_1 = 1 - m_0$; $m_0 = 2/3$, $m_1 = 1/3$. The cascade was continued down to order 8. From Evertsz and Mandelbrot, 1990.

When $\varepsilon \to 0$, the main contribution to the integral comes from values of α close to the minimum of $q\alpha - f(\alpha)$. If $f(\alpha)$ can be differentiated, to get an extremum, it is necessary that:

$$\frac{\partial}{\partial \alpha}\{q\alpha - f(\alpha)\} = 0 \tag{A.44}$$

For a set value of q, this extremum is achieved for $\alpha = \alpha(q)$ with:

$$\frac{\partial}{\partial \alpha} f(\alpha) \bigg|_{\alpha = \alpha(q)} = q \qquad (A.45)$$

And this extremum is a minimum if:

$$\frac{\partial^2}{\partial \alpha^2} f(\alpha) \bigg|_{\alpha = \alpha(q)} < 0 \qquad (A.46)$$

The function $f(\alpha)$ must therefore be convex (Figure A.8). And when reaching the minimum, for $\alpha = \alpha(q)$, the slope of $f(\alpha)$ is q.

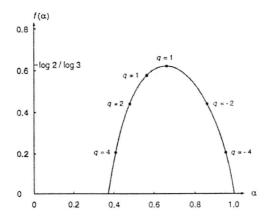

Figure A.8. Multifractal spectrum of the binomial measure defined in Figure A.7. From Evertsz and Mandelbrot, 1990.

Coming back to equation (A.43), and studying the main contribution to the integral, we can define:

$$\tau(q) = q\alpha(q) - f(\alpha(q)) \qquad (A.47)$$

Therefore:

$$\chi_q(\varepsilon) = \varepsilon^{\tau(q)} \qquad (A.48)$$

In the case of binomial measures, we have:

$$\tau(q) = \lim_{k \to \infty} \frac{\log.(m_0^q + m_1^q)^k}{\log.2^{-k}} = -\log_2.(m_0^q + m_1^q) \qquad (A.49)$$

And for multinomial measures:

$$\tau(q) = \log_b \sum_{i=0}^{b-1} m_i^q \qquad (A.50)$$

In equation (A.47), we see that:

$$\frac{\partial}{\partial q}\tau(q) = \alpha(q) \qquad (A.51)$$

$f(\alpha)$ can be computed using $\tau(q)$, and vice versa, because of the relation:

$$f(\alpha(q)) = q\alpha(q) - \tau(q) \qquad (A.52)$$

This relation between $\tau(q)$ and $f(\alpha)$ is a *Legendre transform*.

It can be remarked as well that the function τ can be linked to to the expression of the generalised dimension D_q, defined in Section 1.6 as: $\tau(q) = (q - 1) D_q$.

The practical computation of $f(\alpha)$ by the moments method proceeds along these stages:

1. The measure is coarsely covered by boxes of size ε, i.e. $\{B_i(\varepsilon)\}_{i=1}^{N(\varepsilon)}$ and we determine the measures by box $\mu_i = \mu(B_i(\varepsilon))$.

2. The allotment function from equation (A.40) is computed for different values of ε.

3. One examines if the points $\log.\chi_q(\varepsilon)$ are aligned on the graph as a function of $\log.\varepsilon$. If this is the case, $\tau(q)$ is the slope of the line corresponding to the exponent q (equation A.47).

4. $f(\alpha)$ is computed by Legendre transform of $\tau(q)$ (equation A.52).

These different stages are achieved numerically, because the analytical expressions are generally not calculable (contrary to what Falconer (1990) could achieve for the triadic multifractal set described in Section 1.6).

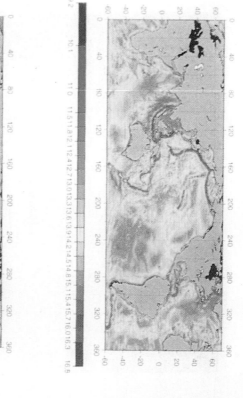

Geoid Roughness

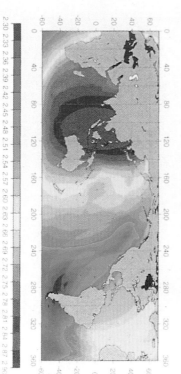

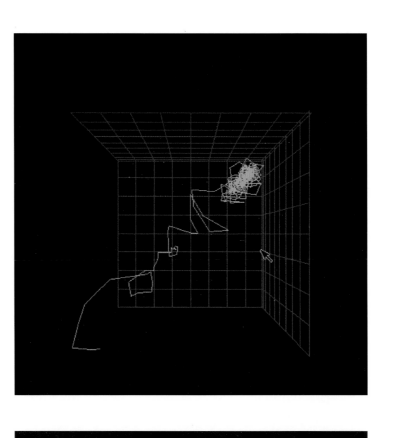

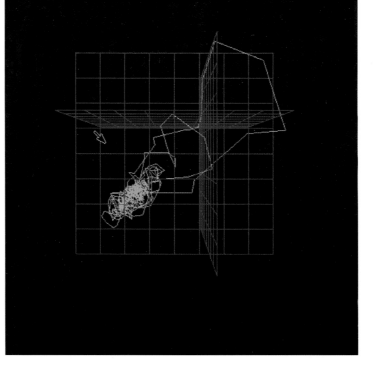

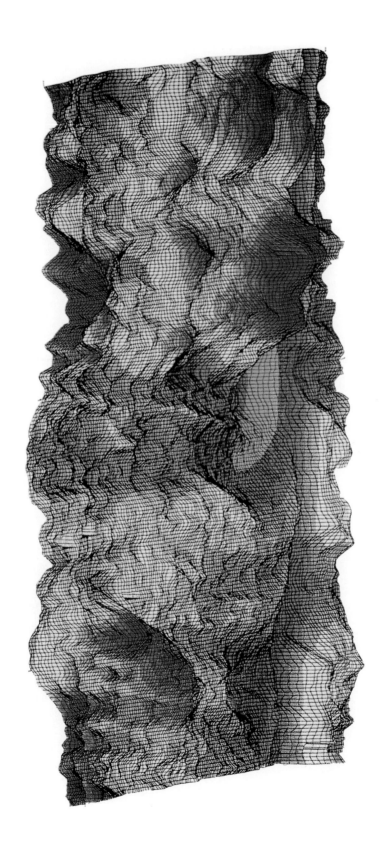

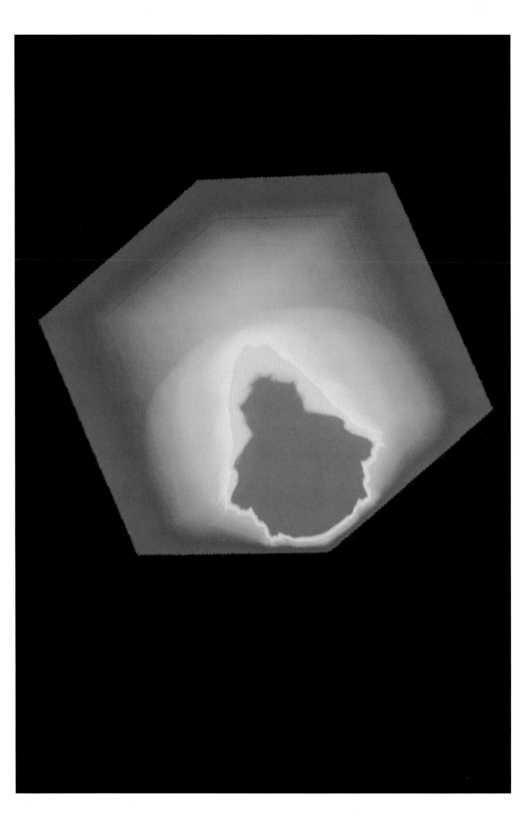

Annex B - Applications to Geophysics

B.1 GEOMORPHOLOGY

B.1.1 Fractal Analysis of River Networks

The results presented here were obtained by Beauvais et al. (1991, 1994) during the investigation of river networks in a continental craton from Central Africa. The study area is located in the Central African Republic, in the region of Zemio, very close to the African continent's centre. This region is drained by the High Mbomou hydrographic network, an affluent of the Congo River (Figure B.1).

The objective was to relate the fractal geometry of river networks to different parameters from the natural environment: along-river slopes, hydrologic parameters, average rainfall, drainage and nature of the underlying rocks.

Richardson's traditional method (Richardson, 1961) was applied. The length $L(\varepsilon)$ of a river is measured with a measuring interval ε, and $N(\varepsilon)$ steps are necessary to cover the whole length: $L(\varepsilon) = N(\varepsilon) \times \varepsilon$.

If the successive lengths L_i are related by a power law to the successive values of the measuring interval ε_i, $1 < i < n$, the river track is fractal with dimension D.

$$N_i = \varepsilon_i^{-D} \text{ and } L_i = \varepsilon_i^{1-D}, \text{ i.e. } D = \lim_{\varepsilon_i \to 0} \frac{\log N_i}{\log(1/\varepsilon_i)} \text{ and } 1 - D = \frac{\log L_i}{\log \varepsilon_i}$$

If the points of the graph of $\log L_i$ vs. $\log \varepsilon_i$ are aligned, the river is fractal and the negative slope of the regression line equals $1 - D$ ($1 < D < 2$).

The application of this technique to the different rivers in the Mbomou Basin produced two types of profiles, with rivers whose along-profile slope is close to 0.6 m/km, and rivers with slopes close to 0.2 m/km (Figure B.2).

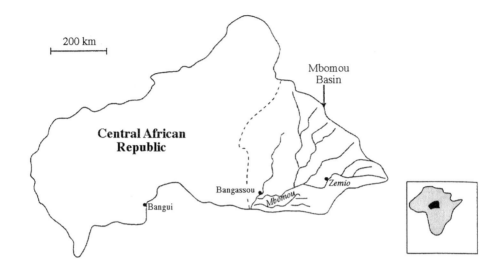

Figure B.1. Geomorphology of a hydrographic network, in the High Mbomou Basin, a tributary of the Oubangui and Congo rivers.

The main results were:

- The fractal dimensions of the studied rivers range between 1.08 and 1.32. They are similar to the dimensions of shorelines.

- Two types of rivers exist, with two power laws depending on the scale considered, i.e. the same scaling law:

    The first group corresponds to rivers with a slope greater than or equal to 0.50 m/km. They are described with two fractal dimensions D_1 and D_2 (Figure B.2). D_1 ranges between 1.16 and 1.32, whereas D_2 ranges between 1.08 and 1.13.

    The second group encompasses all rivers with slopes smaller than or equal to 0.50 m/km. Their single fractal dimension D_1 ranges between 1.13 and 1.29.

- Rivers with the larger slopes generally exhibit narrower valleys and are less sinuous. The dimension D_2 describes the less irregular, narrower courses with larger slopes.

- The dimension D_1 is characteristic of the more irregular and sinuous rivers, with lesser slopes inside larger valleys. The slope has a small influence, contrarily to D_2.

- The dimension D_1 distinguishes the first type of irregularity, unique to rivers and parts of rivers flowing over basic rocks. These rocks (amphibolites, amphiboloschists, dolerites, ...) are highly altered, and better distributed downstream than upstream, where the valleys are large and with small slopes. These zones are also better watered, and vegetation is denser than upstream. The sinuosity is higher, along with sediment flow and transport. Dimension D_1 is called *textural*, as it describes the small-radius sinuosity, which depends from local environmental and geological factors.

- The dimension D_2 distinguishes the second type of irregularity, for rivers or parts thereof which are upstream from the basin and whose slopes are higher. Upstream, these rivers flow over rocks rather more acid rocks (micaschists, gneiss, quartzite). Rainfall is less important, vegetation less dense, and lateritic soils shallower than downstream. The influence of tectonic fracturation is therefore better recorded through sinuosity with large radii of curvature. This is expressed by a smaller fractal dimension, as was observed earlier in the study of fragments from the San Andreas Fault (see Section 2.2). Large-radius sinuosity is mainly controlled by the structure which governs the flow direction. For these reasons, the dimension D_2 is called *structural*.

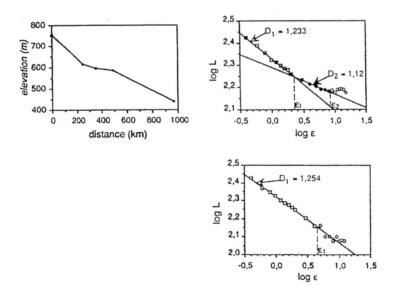

Figure B.2. Fractal dimensions and hydrographic profiles. Rivers in the Mbomou Basin present two types of irregularities with slopes of 0.61 m/km, and one type of irregularity for the Chinko, with a slope of 0.23 m/km. From Beauvais et al., 1994.

To conclude this fractal morphometric analysis of the Mbomou Basin, the course of rivers can be assimilated to a dynamic system with two degrees of freedom, suggested by the two types of irregularities brought into evidence here.

B.1.2 Scale Invariance and Geoid Roughness

Hentschel and Procaccia (1983) defined the family of generalised fractal dimensions D_q so that: $D_q \geq D_{q'}$, $0 \leq q \leq q' \leq \infty$ (see Section 1.6).

The method presented here was proposed by Dubuc et al. (1989). It was adapted to the constraints introduced by the study of the altimetric geoid by Panteleyev et al. (1994) and Panteleyev and Dubois (1994).

$f(\mathbf{t}_i)_{i=1}^N$ is a series of N points arbitrarily chosen from a scalar function $f(\mathbf{t})$, where $\mathbf{t}$ is a parameter vector with (n - 1) dimensions, describing a large number of points in a space with n dimensions ((n - 1) dimensions for the points, and 1 dimension for the value of the function). Let us now cover this space with a grid made from boxes with n dimensions, b_ε^n: $\varepsilon r_1 \times \varepsilon r_2 \times ... \times \varepsilon r_n$, where $\varepsilon \geq 0$ and $1/\varepsilon$ is the number of boxes per unit length on each axis.

The different axes may represent different physical quantities, and each side is multiplied by ε to ensure that the boxes are self-similar at different scales, if possible continuously, and that there is uniform convergence toward zero.

Choosing N as high as necessary, and if $M(b_\varepsilon^n)$ is the number of boxes b_ε^n including N selected points and covering the function $f(\mathbf{t})$ when N tends towards ∞, then:

$$D_0 = \lim_{\varepsilon \to 0} \lim_{N \to \infty} \left[\frac{\log.(M(b_\varepsilon^n))}{\log.(1/\varepsilon)} \right] \tag{B.1}$$

This represents the well-known box-counting algorithm (Turcotte, 1992), and corresponds to the similarity dimension, or first fractal dimension, of Mandelbrot (1967). The methods for computing the fractal dimensions of profiles and surfaces (i.e. functions continuous nearly everywhere) have been discussed by Dubuc et al. (1989). They showed that the box-counting method had a low accuracy, because the function is covered only by an integer number of (square) pixels, giving rise to dramatic jumps when their number decreases.

Minkowski-Bouligand's method of covering with circles is close to the definition of fractals by Mandelbrot, but also produces results with a low accuracy. Two new methods have been proposed. The first one uses horizontal elements and is a progress over Minkowski-Bouligand's method as it requires less computation. The second method, called the variation method, was declared the best because of its rapid convergence toward reliable results, and its smaller computation time.

This method is based on the optimal covering of the function by quadrangles, and gives an estimate of D_0. A direct application of this technique is not possible if the function is defined over a multitude of points with a complex topology. For example, it is impossible to find self-similarity between boxes of different scales if the function is defined on a sphere. Because we want to study global geophysical datasets, we must develop a new technique to analyse functions defined on reduced sets (e.g. 1-D contours or 2-D surfaces). Furthermore, we must retrieve characteristic properties of the function and suppress the influence of edge effects.

For any topology, even with curvilinear box-counting, a suitable measure can be introduced to define the volume of boxes and the volume V of the covering, in order that the boxes have a uniform characteristic volume $\varepsilon^n \times$ (unit measure). The number of covering boxes is V/ε^n.

$$D_0 = \lim_{\varepsilon \to 0} \left[\frac{\log.(V/\varepsilon^n)}{\log.(1/\varepsilon)} \right] \tag{B.2}$$

In a Cartesian topology, equations (B.1) and (B.2) are equivalent if $V = M(b_\varepsilon^n)\varepsilon^n \prod r_i$.

Let $\Gamma(t)$ be a finite multitude, $\mathbf{t}$ a parameter vector with (n - 1) dimensions such that:

$$\exists \mathbf{T}_i, i = 1, 2, ..., n - 1; \sum c_i \mathbf{T}_i \neq 0 \ \bar{V} \ c_i \colon \Gamma(\mathbf{t}) \equiv \Gamma(\mathbf{t} + \mathbf{T}_i) \tag{B.3}$$

Designating with T the largest distance between the points of Γ measured along the shortest line γ between points $\mathbf{t}$ and $\mathbf{s}$ of Γ, we have:

$$T = \max_{\mathbf{t};\mathbf{s} \in \Gamma} \min_{\gamma \subset \Gamma} |\gamma(\mathbf{t}\,;\mathbf{s})| \tag{B.4}$$

$f(\mathbf{t})$ is a continuous scalar function defined on Γ; $f(\mathbf{t})$ describes a multitude of points in an n-dimensional space. Because of Γ's property, $f(\mathbf{t}) \equiv f(\mathbf{t} + \mathbf{T}_i)$. Following Dubuc et al. (1989), we can define the local thickness $v(f,\varepsilon,\mathbf{t})$ of a covering of $f(\mathbf{t})$, and the volume $V(f,\varepsilon,\Omega)$ of the covering of its subset ($\Omega \subseteq \Gamma$), as follows:

$$\alpha(f,\varepsilon,\mathbf{t}) = \sup_{\mathbf{t}' \in \mathcal{R}\varepsilon(\mathbf{t})} f'(\mathbf{t}) \ ; \ b(f,\varepsilon,\mathbf{t}) = \inf_{\mathbf{t}' \in \mathcal{R}\varepsilon(\mathbf{t})} f'(\mathbf{t}) \tag{B.5}$$

$$R_\varepsilon(\mathbf{t}) = \left\{ \mathbf{s}; \min_{\mathbf{t};\mathbf{s} \in \Gamma; \supseteq \gamma \subset \Gamma} |\gamma(\mathbf{t}\,;\mathbf{s})| \leq \varepsilon T; \ 0 < \varepsilon \leq 1 \right\} \tag{B.6}$$

$$v(f,\varepsilon,\mathbf{t}) = \alpha(f,\varepsilon,\mathbf{t}) - b(f,\varepsilon,\mathbf{t}) \tag{B.7}$$

$$V(f,\varepsilon,\Omega) = \int_{\Omega \subseteq \Gamma} v(f,\varepsilon,\mathbf{t}) d\Omega \tag{B.8}$$

As $f(\mathbf{t})$ is supposed to be continuous, so is $v(f,\varepsilon,\mathbf{t})$. Therefore, $v(f,\varepsilon,\mathbf{t})$ can be as small as necessary if $\varepsilon > 0$ is small enough. Consequently:

$$\lim_{\varepsilon \to 0} V(f,\varepsilon,\Omega) = 0 \tag{B.9}$$

As in equation (B.2), the dimension of $f(\mathbf{t})$ can be defined with:

$$D(f,\Omega) = \lim_{\varepsilon \to 0} \left[\frac{\log.(V(f,\varepsilon,\Omega)/\varepsilon^n)}{\log.(1/\varepsilon)} \right] \quad (B.10)$$

Figure B.3. Application of the technique to a function defined on a spherical surface $\Lambda(n=3)$. $W_\varepsilon(t)$ is a moving window of varying size, defined as the isotropic segment of a sphere of radius $\pi\varepsilon$ around a point t, and giving the local thickness $v(f,\varepsilon,t)$ of a covering of the subset Ω of Λ. The moving window $W_\varepsilon(t)$ may encompass surfaces of Λ outside from Ω to avoid edge effects (Panteleyev and Dubois, 1994).

The characteristics of f(t) may vary along the different subsets Ω of Γ, and it is interesting to determine the local singularities of f(t) when Ω tends toward zero:

$$D(f,\mathbf{t}) = \lim_{\Omega \to 0} V(f, \Omega \ni \mathbf{t}) = n - \lim_{\varepsilon \to 0} \left[\frac{\log.(v(f,\varepsilon,\mathbf{t}))}{\log.(\varepsilon)} \right] \quad (B.11)$$

because $\varepsilon \to 0$ for a particular, constant Ω, and thus:

$$\lim_{\Omega \to 0} \lim_{\varepsilon \to 0} \frac{\log.(V(f,\varepsilon,\Omega))}{\log.(\varepsilon)} = \lim_{\Omega \to 0} \left(\lim_{\varepsilon \to 0} \frac{\log.(v(f,\varepsilon,t))_\Omega}{\log.(\varepsilon)} + \lim_{\varepsilon \to 0} \frac{\log.(\Omega)}{\log.(\varepsilon)} \right)$$

$$= \lim_{\varepsilon \to 0} \left[\frac{\log.(v(f,\varepsilon,\mathbf{t}))}{\log.(\varepsilon)} \right] \quad (B.12)$$

and:

$$D(f,\mathbf{t}) = n - \lim_{\varepsilon \to 0} \left[\frac{\log.(v(f,\varepsilon,\mathbf{t}))}{\log.(\varepsilon)} \right] \quad (B.13)$$

Let us remark that $(v(f,\varepsilon,t))_\Omega$ is the mean value of $v(f,\varepsilon,t)$ in Ω. The definition (B.12) is always applicable, provided that $v(f,\varepsilon,t)$ remains strictly positive.

This definition of $D(f,t)$ allows to map the invariant features characteristic from the terrain. This method differs from the classic filtering method, because it highlights the main features at different scales, whereas the filtering method investigates only a frequency range defined *a priori*.

Log-log graphs of the points $\begin{Bmatrix} -\log.(\varepsilon) \\ \log.(v(f,\varepsilon,t)/\varepsilon^n) \end{Bmatrix}$ and $\begin{Bmatrix} -\log.(\varepsilon) \\ \log.(V(f,\varepsilon,\Omega)/\varepsilon^n) \end{Bmatrix}$ show slopes which yield the respective dimensions $D(f,t)$ and $D(f,\Omega)$.

Finally, the roughness functions $R(f,t)$ and $R(f,\Omega)$ are introduced as the intersections of the log-log plot with the reference mean logarithmic scale $\varepsilon_m = (\varepsilon_{min} - \varepsilon_{max})^{1/2}$, where ε_{min} and ε_{max} are the limits of linear scale invariance:

$$\{\log.(V(f,\varepsilon,\Omega)/\varepsilon^n)\}_{lin} = D(f,\Omega) \log. (\varepsilon/\varepsilon_m) + R(f,\Omega)$$

$$\{\log.(v(f,\varepsilon,t)/\varepsilon^n)\}_{lin} = D(f,t) \log. (\varepsilon/\varepsilon_m) + R(f,t) \qquad (B.14)$$

The index $_{lin}$ means that the values are linearised. The higher the amplitude of the signal, the higher the value of R.

This definition of roughness in terms of scale invariance differs from the one of Malinverno and Cowie (1993), which is the square root of variance, and from the one of Turcotte (1992), which is the intercept of a linear power spectrum, as well as from the filtered envelope surface of a short-wavelength interval (Goslin and Gibert, 1990).

Data and results

The altimetric data used for this application come from the Geosat Mission; they are gridded with a spatial resolution of 7.5' × 7.5', i.e. 14 × 14 km² at the equator and denser elsewhere, and their rms precision is of approximately 10 cm along-track (Houry et al., 1993). The region covered extends from 60°S to 70°N. This represents a file of 2,053,801 measure points spread over 63.8% of the Earth's surface. The zones not covered are the continents and coastal areas (26.5%) and the ice caps (9.7%). Compared to the reference ellipsoid, the geoid varies from -105.44 m to + 81.54 m. The study domain therefore spans from 30 km to 40 10^3 km, around 3 orders of magnitude.

Admitting that the global altimetric geoid is a function defined on a sphere, the computation with a moving window of increasing size produces logarithmic scales quasi-uniform from 15 km to 20 10^3 km in radius (Figure B.4).

The preliminary interpretation of these first results can related the wavelength domains to the depth of sources and to density heterogeneities. Thus, the high value of the long-wavelength dimension (2.5) may be related to the high roughness of the core-mantle boundary. The low value of 2.1 (third line segment in the log-log graph) may be characteristic from the upper mantle and express a regular stratification. And the intermediate value of 2.2 (second line segment in the log-log graph) and the associated high dispersion may correspond to the irregularities in the deep mantle (convection ?).

The lithosphere is characterised by a regional variability which translates well its heterogeneity. Thus, fracture zones and subduction zones show relatively high dimensions (at short wavelengths, 30 to 150 km). Conversely, abyssal plains and mid-ocean ridges exhibit low dimensions (lithospheres with a small roughness). Colour Plate 3 (between pp. 192 and 193) further illustrates these variations.

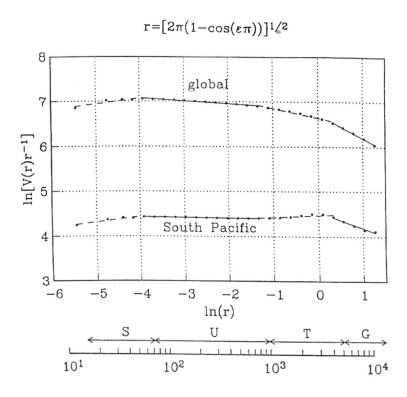

Figure B.4. Scale invariance of the altimetric geoid. A global and a regional analysis respectively highlight four and three domains with different fractal dimensions. This implies the variations of structures depending on the investigation depth (depending on the wavelength).

B.2 FRAGMENTATION, FRACTURATION, TECTONICS AND PERCOLATION

B.2.1 Fracturation and an Anisotropic Model

The model of Allègre and Le Mouël

The model of brittle rock fracturation described here was proposed by Allègre and Le Mouël (1994) and complements the model presented in Section 2.2. Contrary to the previous one, this model is anisotropic with normal and tangential fracture probabilities, interacting between each other.

We consider a tectonic domain with a stress field $\sigma_1 \geq \sigma_2 \geq \sigma_3$, and as first approximation we will use the 2-D space in which the stress tensor is represented by its two principal components σ_1 and σ_3.

As can be seen on Figure B.5, each fundamental domain of order 1 is formed of 3×3 domains of order 2, which are themselves formed of 3×3 domains of order 3, etc. At any scale, a domain has the probability p of becoming fragile, and therefore a probability $(1 - p)$ of remaining sound. But, contrary to the model presented in Section 2.2, the probability in the principal direction p_M on σ_1 is considered different from the probability in the secondary direction p_S on σ_3.

These conditions are not independent but coupled, and the coupling obeys to rules derived from field observations and laboratory experiments. Figure B.6 shows the interaction of the central square with two possible types of simple and complex interactions.

The first, simple interaction is an application of the stress distribution around a fracture, according to the elasticity theory and observations. The second, complex interaction is based on the geological and experimental observation known as *en-échelon* interaction.

The first is translated by the probability for the central square to become brittle by interaction when it was not brittle before (probability $(1 - p)$). The interaction with the neighbouring domains generates an additional breaking probability, which is proportional to $(1 - p_M)$ or $(1 - p_S)$ as well as p_M and p_S.

For the second interaction, if the central square is brittle (probability p), the presence of the neighbouring domains reduces the breaking probability proportionally to p_M or p_S multiplied by another function of p_M and p_S.

This situation is not the most usual, and can be formulated as follows:

$$\begin{cases} \pi_{M;1} = p_M + (1 - p_M)(a_M p_M + d_M p_S) - p_M(b_M p_M + c_M p_S) \\ \pi_{S;1} = p_S + (1 - p_S)(a_S p_S + d_S p_M) - p_S(b_S p_S + c_S p_M) \end{cases} \quad (B.15)$$

This can be simplified as:

$$\begin{cases} \pi_{M;1} = f_M(p_M; p_S) \\ \pi_{S;1} = f_S(p_S; p_M) \end{cases} \quad (B.16)$$

In the first expression, $\pi_{M;1}$ and $\pi_{S;1}$ are the principal and secondary probabilities for a domain of order 1 to be brittle, and are functions of the probabilities p_M and p_S and the interactions with the neighbouring domains. The coefficients a_M, b_M, ... and d_S are chosen so that $0 \leq \pi_{M;1}, \pi_{S;1} \leq 1$ if $0 \leq p_M, p_S \leq 1$.

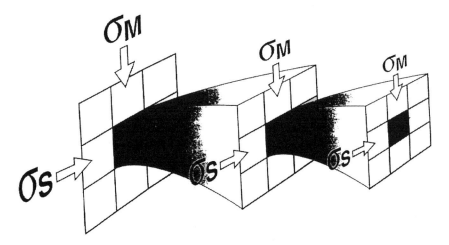

Figure B.5. The scaling law in the model of Allègre and Le Mouël (1994). A grid of 9 squares is reproduced inside each square at each new stage.

To pass from the order 1 to the order 2, we assume that a square formed of 3×3 domains is brittle in the normal direction if it contains at least 3 brittle squares aligned along the principal direction. To be brittle in the secondary direction, the 3 squares need to be aligned along the secondary direction.

This scaling relation is expressed by a polynomial relation, like the model described in Section 2.2:

$$\begin{cases} p_{M;2} = G(\pi_{M;1}) \\ p_{S;2} = G(\pi_{S;1}) \end{cases} \quad (B.17)$$

where $G(x) = 3\,x^3\,(1-x)^6 + 18\,x^4\,(1-x)^5 + 45\,x^5\,(1-x)^4 + 57\,x^6\,(1-x)^3 + 36\,x^7\,(1-x)^2 + 9\,x^8\,(1-x) + x^9$. This relation defines the addition of several breaking geometries.

We can also define a composite probability at the order 2, as we did at the order 1:

$$\begin{cases} \pi_{M;2} = f_M(p_{M;2};p_{S;2}) \\ \pi_{S;2} = f_S(p_{S;2};p_{M;2}) \end{cases} \quad (B.18)$$

At order k, we will have:

$$\begin{cases} p_{M;k} = G(\pi_{M;k-1}) \\ p_{S;k} = G(\pi_{S;k-1}) \end{cases} \quad (B.19)$$

$$\begin{cases} \pi_{M;k} = f_M(p_{M;k};p_{S;k}) \\ \pi_{S;k} = f_S(p_{S;k};p_{M;k}) \end{cases}$$

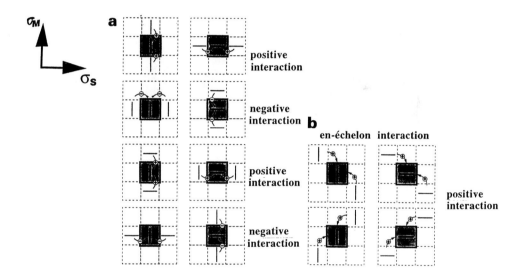

Figure B.6. Breaking probabilities of a central square as a function of its environment. Two types of interaction are considered: (a) in the simple interaction, the stress distribution around the fracture follows the theory of elasticity; (b) in complex interactions, geological and experimental observations are refered to (e.g. *en-échelon* interactions).

A number of assumptions are made to establish the fracture behaviour model.

$p_{M;1}$ and $p_{S;1}$ are considered to be dependent on the difference $(\sigma_1 - \sigma_3)$.

If the difference $(\sigma_1 - \sigma_3)$ increases with time, $p_{M;1}$ and $p_{S;1}$ will also increase with time. This variation is assumed to be linear.

We assume that the ratio $\alpha = (p_{M;1}(t) / p_{S;1}(t))$ remains constant. This means that the deformation anisotropy remains constant through time.

Two types of models can then be considered, corresponding to two types of interaction between the normal and tangential probabilities:

- negative interactions: the decrease in the breaking probability in one direction results from the increase in the other direction. This self-healing case corresponds to $(c_M + d_M, c_S + d_S > 0)$.

- positive interactions: the increase of probability in one direction corresponds to an increase in the other direction. There is a kind of cooperation between directions, and $(c_M + d_M, c_S + d_S < 0)$.

We assume that, at all levels, the breaking probabilities $p_M(t)$ and $p_S(t)$ are combining as vectors. The number N_k of fractures at level k is proportional to the amplitude of the vector sum of $p_M(t)$ and $p_S(t)$:

$$P_k(t) = \frac{\sqrt{2}}{2} \sqrt{[p_{M;k}(t)]^2 + [p_{S;k}(t)]^2} \tag{B.20}$$

and $N_k \propto P_k$.

The orientation of fractures is given by:

$$\theta_k(t) = \tan^{-1}\left\{\frac{p_{M;k}(t)}{p_{S;k}(t)}\right\} \tag{B.21}$$

For the order 1, we get from the definition of α: $\theta_1 = \tan_{-1}.\alpha$

Numerical results

The model normally depends on 8 parameters, but because of our assumptions, their number was reduced to 4 in each of the two cases considered.

The self-healing case

In this case, the interaction between the principal and secondary probabilities is negative. With $d_M = b_M = d_S = b_S = 0$, we have:

$$\begin{cases} \pi_M(t) = p_M(t) - c_M p_M(t) p_S(t) + a_M(1 - p_M(t))p_M(t) \\ \pi_S(t) = p_S(t) - c_S p_S(t) p_M(t) + a_S(1 - p_S(t))p_S(t) \end{cases} \tag{B.22}$$

The fact that $c_M, c_S, a_M, a_S \geq 0$ expresses the mutual influence of the different types of brittleness, related to the elastic approximation for the non-diagonal terms of the stress tensor. To reduce the number of parameters and introduce convenient symmetries, we choose $a_M = a_S = 0.5$; $c_M = c_S = 0.5$. And $\alpha = (p_{M;1}(t) / p_{S;1}(t))$ is chosen between 1 and ∞, so that the values of θ_1 range between 45° and 90°.

Numerical simulations show that the model's behaviour strongly depends on the numerical value of α.

- For $1 < \alpha < \alpha_c$, $\tan^{-1}.\alpha_c \approx 52°$, there never appears any catastrophic behaviour. There are, however, small fractures which may appear at several scales. This behaviour is called creep, and is observed by seismologists only during small earthquakes. Geologists talk about plastic deformation. The environment only produces small-scale fractures.

- For $\alpha > \alpha_c$, the situation is very close to the isotropic case mentioned earlier. For small values of k, $p_{M;k}(t)$ increases slowly and regularly. For high values of k, the critical process occurs; from this point, the breaking probability is the same at all levels. The catastrophic behaviour is not preceded (announced) by premonitory events of a smaller order. When the two probabilities $p_M(t)$ and $p_S(t)$ are of the same scale, the organisation disappears and is replaced by creep. If one is much bigger than the other, this goes back to the 1-D case. The free parameters can be changed, as α_c, and the creep domain can be extended, while the fragile ductile transition can be determined as a function of these parameters.

Figure B.7 illustrates the properties of this model.

The cooperating case

Let us now consider the case where $p_M(t)$ and $p_S(t)$ are cooperating positively, i.e. $a_M = b_M = c_M = a_S = b_S = c_S = 0$.

$$\begin{cases} \pi_M(t) = p_M(t) + d_M(1 - p_M(t))p_S(t) \\ \pi_S(t) = p_S(t) + d_S(1 - p_S(t))p_M(t) \end{cases} \tag{B.23}$$

We look at the case where $d_M = d_S = 0.3$; $d_M = 0.7$ and $d_S = 0.3$.

In this case, the behaviour is always catastrophic and critical for a few values of p(t).

Fracture geometry

The values of ω_k provide the orientations of fractures at the order k. Figure B.8 sums up the results obtained in the different cases. It shows how the fractures are oriented at different orders, in the self-healing case and in the cooperating case. These two situations can be observed experimentally. There are two possible cases for each: when the confining pressure is low, the fractures at different scales are oriented along σ_1; when the confining pressure is increased, the small fractures are still oriented along σ_1 but macroscopic fractures appear at 45° from σ_1.

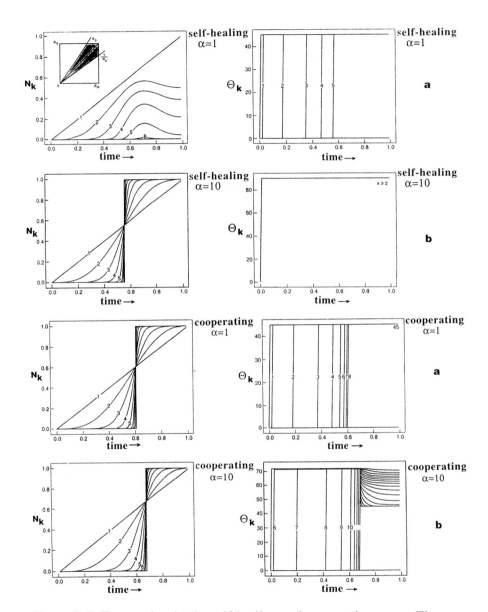

Figure B.7. Fracturation in the self-healing and cooperating cases. The number of fractures N_k and the breaking angle θ_k are represented for different orders $k = 1, 2, ..., n$, below and above the critical angle α_c.

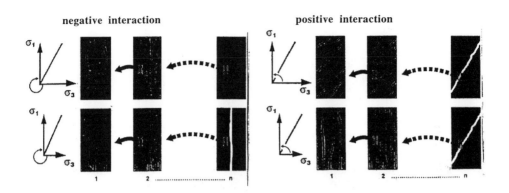

Figure B.8. The orientation of fractures is shown at several levels for the two cases considered, and with a low or high confining pressure (see text for details).

The SOFT model (Allègre et al., 1995)

The acronym SOFT stands for "Self-Organised Fracturation and Tectonics". Contrary to the previous models (previous section and Section 2.2), Allègre et al. (1995) account for the system's energy budget. Energy is regularly brought by the process of global plate tectonics, and dissipated as earthquakes. The underlaying fracturation process is still based on the technique of renormalisation groups used in the previous models.

Without entering into the computations, we will show the main result from this model. The behaviours observed will depend on the amount of energy provided to the system. Numerical simulations show that three cases can occur:

- If the energy provided slowly approaches a critical threshold of energy per surface unit, the system answers by losing energy continuously without any critical phenomenon.

- If the rate of energy provided is higher, premonitory breaks occur (pseudo-critical phenomena), but without releasing the equivalent energy, before reaching a genuine critical point where a large amount of energy is released in a large earthquake.

- For a higher increase in the energy provided, the system abruptly reaches the critical point and a large earthquake happens, without precursors.

Figure B.9 illustrates these three cases. Let us remark that the sum of the different experiments is a good simulation of the Gutenberg-Richter law. It would now be interesting to test these different series of events with the techniques presented in this book (Cantor dust, correlation function, etc.). The similarity of this model with the sand-heap model is indeed important.

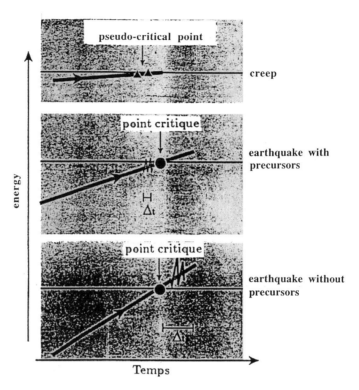

Figure B.9. The SOFT model uses renormalisation groups and the balance between the energy provided by global plate tectonics and the energy released by earthquakes. The three possible situations are shown here, and detailed in the text. From Allègre et al., 1995.

B.2.2 Percolation, a Geometric Disorder

The following developments aim at providing bases for the study of percolation networks in application to magma chambers, water tables, or hydrocarbon reservoirs.

This perspective was adopted by Roux (1990), and consists in investigating a situation where a simple critical point of geometrical transition occurs; percolation. We will start by recapitulating a few geometrical properties.

Let us consider a lattice of infinite size, and cut randomly a fraction (1 - p) of the lattice's nodes. If p is close to 1, there are only a few holes in the lattice. If p is close to 0, the lattice is divided into small clusters.

Between these two extremes, the sizes of clusters are widely distributed, and there exists a critical value p_c, called the *percolation threshold*, such that, if $p > p_c$, there is a cluster of infinite size (transportation through the lattice is then possible), and if $p < p_c$, there are only clusters of finite sizes.

The objective of percolation studies is to describe the statistical properties close to the percolation threshold.

This physical process has all the characteristics of a second-order phase transition, the fraction p being the control parameter. The order parameter is the possibility of belonging to the infinite cluster.

There are two types of properties, both linked: those which are geometrical and those that deal with transportation processes.

Percolation being a critical phenomenon, its properties will have a universal character, i.e. they will not depend on the fine-scale details of the structures we will study.

Lattices

Figure B.10 shows the different types of bi-dimensional lattices: square, triangular, hexagonal, Kagomé (Essam, 1980).

Table B.1. Values of the percolation threshold for several lattice geometries.

Lattice	Site Percolation	Node Percolation
2-D Lattices		
Square	0.59273	1/2
Triangular	1/2	$2\sin(\pi/18) = 0.3473$
Hexagonal	0.6962	$1 - 2\sin(\pi/18) = 0.6527$
Kagomé	$1 - 2\sin(\pi/18) = 0.6527$	
3-D Lattices		
Cubic simple	0.3116	0.2488
Cubic centred	0.245	0.1785
Cubic with centred sides	0.198	0.119
Diamond	0.428	0.388
4-D Lattices		
Hypercubic simple	0.198	
5-D Lattices		
Hypercubic simple	0.142	

The percolation threshold will vary depending on the geometry of the lattice and the way it has been degraded. The nodes of a hexagonal lattice can for example be removed randomly, as in Figure B.11, where the fraction p is 1 - 0.653, which is the percolation threshold. Here, we have a *node percolation*. It would have been also possible to degrade the lattice by removing sites, and reach a *site percolation threshold*. The percolation thresholds differ, for a particular type of lattice, between these two types of degradation (cf. Table B.1).

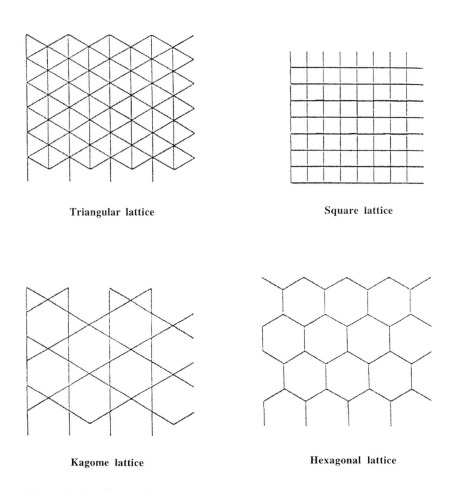

Figure B.10. These four basic lattices have node percolation and site percolation thresholds, which can be computed formally or numerically.

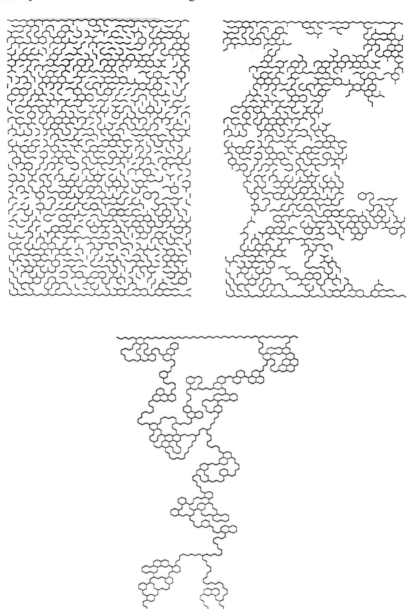

Figure B.11. Hexagonal lattice, degraded by randomly cutting a fraction 1 - 0.653 of the nodes, yielding the node percolation threshold. The second figure corresponds to a fractal object of dimension 1.89, and the third one to a fractal object of dimension 1.62. From Roux, 1990.

Critical exponents

Let us go back to the notion of infinite cluster defined in Section 2.2. In the case of a lattice of finite size, if a cluster intersects the lattice (cf. Figure B.11), the lattice is considered as infinite. This is like the definition of the archipelago enabling the crossing between two continents far away, but at a finite distance from each other. Several critical exponents come into play.

(1) The exponent β. We saw that, for percolation, the order parameter was the probability to belong to an infinite cluster P_∞. For $p < p_c$, $P_\infty = 0$. For $p > p_c$, $P_\infty \neq 0$ and becomes null at the threshold with the power law:

$$P_\infty \approx (p - p_c)^\beta \tag{B.24}$$

The value of β only depends on the space's dimension (see Table B.1).

(2) The exponent τ and the size distribution of clusters. The number of clusters of size s is n_s. At the percolation threshold and for an infinite cluster, the distribution follows a power law:

$$n_s \approx s^{-\tau} \tag{B.25}$$

A consequence is that there is no characteristic cluster size. This is a property of critical processes.

(3) The exponent σ. When moving away from p_c, for a particular value of p, the distribution in (B.25) becomes:

$$n_s = s^{-\tau} f((p - p_c)s^\sigma) \tag{B.26}$$

The numerical study of this distribution shows that there is a unique scaling function f, which can account for results for different values of p. This function looks like a Gaussian. σ is a critical exponent.

(4) Relationship between β, τ and σ. Beyond the percolation threshold, a node belongs either to the infinite cluster, or to a finite cluster. One deduces:

$$P_\infty + \sum sn_s = p \Rightarrow \partial P_\infty/\partial P = 1 - \sum s\partial(n_s)/\partial p \tag{B.27}$$

The sum can be replaced by a continuous integral:

$$\partial P_\infty/\partial p = 1 - \int ss^{-\tau}s^\sigma f((p - p_c)s^\sigma)ds \tag{B.28}$$

If:

$$A = \int z^{(2-\tau)/\sigma} f(z)dz \tag{B.29}$$

Then:

$$\partial P_\infty/\partial p = 1 - (A/\sigma)(p - p_c)^{(\tau-2)/(\sigma-1)} \tag{B.30}$$

And, because of β's definition:

$$\beta = (\tau - 2)/\sigma \tag{B.31}$$

(5) The exponent v. s_{max} is the upper threshold of cluster size. The number of nodes occupied by a cluster of size z is:

$$\xi^d \Delta p^\beta = s_{max} \tag{B.32}$$

and:

$$s_{max} = \Delta p^{-1/\beta} \Rightarrow \xi^d \approx \Delta p^{-1/(\sigma-\beta)} \tag{B.33}$$

that is:

$$\xi = \Delta p^{-\nu}, \text{ where } \nu = (1/\sigma + \beta)/d \tag{B.34}$$

This exponent v is the same one defined in Section 2.2, and controls the length scale. The length ξ is called the correlation length, and we deduce:

$$\sigma = 1/(\nu d - \beta) \text{ and } \tau = 2 + \beta\sigma = 2 + \beta/(\nu d - \beta) \tag{B.35}$$

(6) The exponent γ. The average size χ of clusters below the percolation threshold corresponds to:

$$\chi = \sum s^2 n_s \tag{B.36}$$

Using the same approach as in the previous paragraphs, we get:

$$\chi = \int s^2 s^{-\tau} f((p - p_c)s^\sigma) ds \tag{B.37}$$

and if $B = \int z^{(3-\tau)/\sigma} f(z) dz$, then:

$$\chi \approx (B/\sigma) \Delta p^{(\tau-3)/\sigma} \tag{B.38}$$

χ can be considered to diverge with the exponent γ: $\chi = (\Delta p)^{-\gamma}$, and therefore:

$$\gamma = (3 - \tau)/\sigma \tag{B.39}$$

The analysis can be generalised for all moments m of the distribution (we just saw m = 1 and m = 2). We can write:

$$\sum s^m n_s = (\Delta p)^{\nu(d - mD_f)} \tag{B.40}$$

The exponents (β for m = 1, γ for m = 2, $\nu(d - mD_f)$) are affine in m.

Table B.2 gives the values of critical exponents for spaces with different dimension.

Table B.2. Values of the critical exponents as a function of the space dimension.

Space Dimension	2	3	4	5	6
Geometry					
ν	4/3 = 1.33	0.88	0.64	0.51	1/2
β	5/36 = 0.14	0.44	0.56	0.67	1
γ	43/18 = 2.39	1.76	1.43	1.21	1
τ	187/91 = 2.05	2.20	2.28	2.35	5/2
σ	36/91 = 0.39	0.45	0.50	0.53	1/2
D_f	91/48 = 1.89	2.50	3.12	3.69	4
D_{bb}	1.62	1.74	1.90	1.93	2
D_{min}	1.13	1.34	1.50	1.75	2
D_{scb}	3/4 = 0.75	1.14	1.56	1.96	2
Transportation					
D_s	1.32	1.33	1.35	1.35	4/3
D_w	2.87	3.77	4.64	5.45	6
t	1.30	2.02	2.25	2.43	0
s	1.30	0.75	0.60		0
τ	3.96	3.8			4
σ	1.30				

<u>Finite-size lattices</u>

The parameter just defined are used to quantify situations where the percolation threshold has not exactly been reached. Reasoning now on a lattice of finite size L, we can introduce a scale function φ:

$$P_\infty = (p - p_c)^\beta \, \varphi(L/\xi) \tag{B.41}$$

In his study, Roux (1990) postulates that at the neighbourhood of 0, $\varphi(x)$ can be expressed as $\varphi(x) = x^\zeta$, where ζ is still unknown.

As $\xi \approx (\Delta p)^{-\nu}$, we have: $P_\infty \approx (\Delta p)^{\beta+\zeta\nu}$ and $\zeta = -\beta/\nu$.

Therefore: $\varphi(x) \approx x^{-\beta/\nu}$, $x \ll 1$.

And for a lattice of size L:

$$P_\infty \approx L^{-\beta/\nu} \text{ if } L \ll \xi \tag{B.42}$$

For a lattice of size L at the percolation threshold, the mass of the percolating cluster, if it exists, will be proportional to:

$$L^d P_\infty(L) \approx L^{d-\beta/\nu} \tag{B.43}$$

This means that the infinite cluster is a fractal object of dimension:

$$D_f = d - \beta/\nu \tag{B.44}$$

Schematisation of the infinite cluster

Let us consider an infinite environment at the percolation threshold, with a hexagonal lattice (Figure B.12a). After randomly cutting links up to the percolation threshold, some structures are linked to the main cluster by single terminations only. These structures are called "dead ends". They play no role in the transport processes, except if they are sources.

Figure B.12b shows the largest cluster in the lattice at the percolation threshold. This cluster is fractal, with a dimension of 1.89. On Figure B.12c, part of the infinite cluster has been retained after removing all structures which are not connected to the main cluster through single links ("dead ends"). This structure is called the "skeleton" of the infinite cluster.

Figure B.12d is identical to Figure B.12c, except that the links are represented in bold. They are called "sensitive links", as the lattice is not percolating any more once they are cut. These links form a fractal object of dimension 3/4. Clusters of loops, called blobs, are also visible in this figure. They are linked together by single links, i.e. sensitive links. If they get cut, the skeleton gets disconnected into 2 parts. The sensitive links form a fractal set of dimension $1/\nu$.

The exponent τ is associated to the size distribution $n_s \approx s^{-\tau}$. It is related to the skeleton's fractal dimension by a relation identical to the relation for finite clusters:

$$\tau = 1 + d/D_b b \tag{B.45}$$

The shortest path from one edge to the other defines a new fractal object of dimension D_{min} (see Table B.2). This object is represented on Figure B.13, and is a fractal object of dimension 1.13.

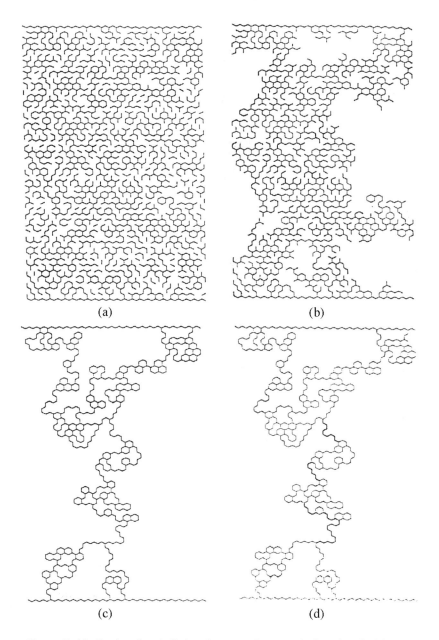

Figure B.12. Study of an infinite cluster at the percolation threshold. Several operations allow the definitions of structures called "dead ends", "skeleton", "sensitive links", "blobs", etc. (see text for detail). These structures have established fractal dimensions. From Roux, 1990.

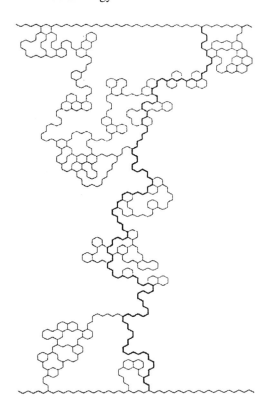

Figure B.13. The shortest path between the extreme edges of the lattice are plotted in bold.

B.3. SEISMOLOGY - RUSSIAN METHODS

B.3.1 Prediction Method of P. Shebalin and I. Rotwain

The two methods M8 and CN were exposed in Section 2.7. They allow the diagnosis of strong earthquakes in large regions with linear dimensions of several hundred kilometres. The duration of TIPs (Times of Increased Probability) of strong earthquakes is of approximately two years. It is therefore natural to look for ways of improving the forecasting by reducing both the geographical area and the duration of the TIPs.

Shebalin et al. (1994) proposed a method where the search for precursors is narrowed to the close neighbourhood of strong earthquakes. Precursors have indeed been found close to the epicentre of an imminent strong earthquake. This was observed by analysing the spatial and temporal distribution of small earthquakes in Southern California, in the framework formed with morphostructural lineaments. Small earthquakes usually cluster along these lineaments, particularly at their intersections (Gelfand et al., 1976). The authors observed that several months before the strong earthquakes, close to their epicentre and on short time scales (1 to 3 months), clustering was interrupted and a relatively large number of earthquakes occured away from the lineaments. The number of small earthquakes increased as well.

The construction of these lineament frameworks is no easy task. Consequently, we shall focus more on the ideas based on the use of seismic catalogues, rather than the lineament maps.

The method

We shall consider the epicentres of small earthquakes with magnitudes $M \geq M_{min}$.

Time distribution

At each instant t, two time intervals are defined:

- a relatively short interval, of duration T_c: $C = [t - T_c; t]$, in which the spatial distribution of epicentres will be investigated.
- a long interval, of duration T_b: $B = [t - T_d - T_b; t - T_d]$, in which the long-term distribution of epicentres will be investigated.

The intervals B and C are separated by the time T_d (Figure B.14).

Spatial distribution

S is the geographic square centred on the point (φ, λ), covering D degrees in latitude and longitude (see Figure B.14). This square is subdivided into two areas.

The area $\mathcal{A}$ is by definition the part of S for which, during the time interval B;
  (1) the number $N_{\mathcal{A}}$ of epicentres in $\mathcal{A}$ is the portion q of the total number N of epicentres in the square S.
  (2) the number of epicentres in the square box of size d, centred anywhere in $\mathcal{A}$, is greater than the corresponding number for any point of S outside $\mathcal{A}$. This supplementary area is called $\mathcal{N}$ (Figure B.14).

Thus:

$$\mathcal{A} \cup \mathcal{N} = S \qquad (B.46)$$

The number of epicentres in the area $\mathcal{N}$ during the time interval B is $N_{\mathcal{N}}$.

$$N = N_{\mathcal{A}} + N_{\mathcal{N}}; \ N_{\mathcal{A}} = q N \tag{B.47}$$

During the time interval C, the total number of epicentres is n, the number of epicentres in $\mathcal{A}$ is $n_{\mathcal{A}}$ and the number of epicentres in $\mathcal{N}$ is $n_{\mathcal{N}}$.

The effect described earlier will be described by the following function, defined for the square S and at instant t:

$$f(t, \varphi, \lambda) = \frac{n_{\mathcal{N}} \ N_{\mathcal{A}}}{(n_{\mathcal{A}} + 1)(N_{\mathcal{N}} + 1)} \tag{B.48}$$

M_{min}, T_b, T_c, T_d, D, d and q are the free parameters of the definition of f.

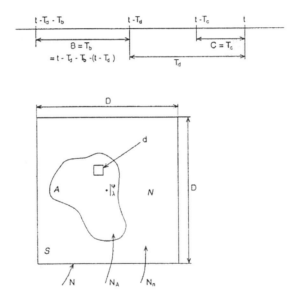

Figure B.14. Space-time method. This figure represents in space and time the main parameters used in the method of Shebalin and Rotwain.

Example of an earthquake in Imperial Valley

Shelbain (1993) uses the example of an earthquake which occurred on October 15, 1979, with parameters (φ = 32°06'N, λ = 115°02'W, M = 6.9). The free parameters used were: M_{min} = 2.5, T_b = 30 years, T_c = 60 days, T_d = 0, D = 0.5, d = 0.07 and q = 0.5. Computations were performed for a square centred on (32°09'N, 115°04'W), for a time interval of 3.5 months before the earthquake. The resulting value of the function is f = 1.64.

Figure B.15 shows the time variations of f at proximity of a few strong earthquakes. The vertical bars indicate the start hour of earthquakes with magnitudes ≥ 5.5 and less than 100 kilometres away from the centre of the square considered. Relatively high values of the function are observed 3 to 12 months before earthquakes of magnitude ≥ 6. Similar values are also observed immediately after earthquakes of magnitude 5.5.

Forecasting algorithm

The function f can be used to build an algorithm of earthquake forecasting. $\mathcal{G}$ is a seismic area in which the catalogue of earthquakes is assumed to be reliable since a date t_0. The values of f can be computed from the time $t = t_0 + T_b + T_c + T_d$.

The problem is equivalent to determining in a space-time domain the occurrence of an earthquake of magnitude $M \geq M_0$.

This domain can be defined with a terminology similar to the one used with the M8 and CN algorithms for TIPs. Here, we will designate the space-time domains as TSIP (Time-Space areas of Increased Probability), or, more simply, "alarms".

$\mathcal{G}$ is covered by a lattice of constant step δ ($\delta < 0$). The square cells S are centred on the nodes of the lattice. Similarly, the time domain is sampled with a constant step τ, and we consider the successive times t_0, $t_1 = t_0 + \tau$, $t_2 = t_0 + 2\tau$, ...

If, at a time t_i and in any square S, the value of f is greater or equal to a given threshold F, an "elementary alarm" is marked.

The union of squares S where alarms are observed forms the area of this alarm at time t_i. The alarm is raised in this area for a time T. The corresponding space-time domain (alarm area x time interval T) makes up the elementary TSIP. The union of elementary TSIPs overlapping in time and space makes the TSIP itself.

If a strong earthquake ($M \geq M_0$) effectively occurs during this TSIP, the forecast is correct. This alarm disappears immediately after the earthquake. If there is no earthquake verifying $M \geq M_0$, this is a "false alarm". If the earthquake occurs outside from the TSIPs, there is a "prediction error".

M_0, δ, τ, F, T and the parameters of function f are the free parameters of the algorithm.

The *a posteriori* application of this algorithm to South California produced forecasts correct in 6 cases out of 8, for a total volue of TSIPs being 6.8% of the total space-time domain considered.

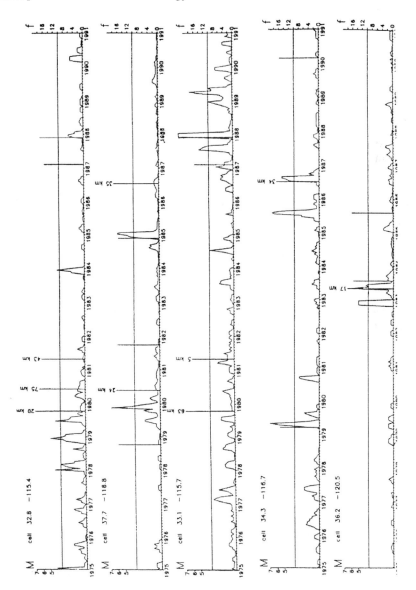

Figure B.15. Time evolution of the function f, for 5 different square cells of a seismic region of California. The vertical bars correspond to the proximity of strong earthquakes. From Shebalin, 1993.

B.4 GEOMAGNETISM

B.4.1 Study of a Long Magnetic Series

The study presented here was performed by Hongre et al. (1994) on a long series of hourly measures of the Earth's magnetic field, recorded at the observatory of Chambon-la-Forêt (France). It consists of 14,500 observations for a period of 1.5 years (1988-1989).

The main variations are related to the varied current systems in the ionosphere and magnetosphere (external part of the field), and to secular variations (internal part of the field). The latter are considered as approximately constant for the short time interval of 5 years. Because we are investigating a short time series, the main causes of field variations to be accessible will be the external ones. Field variations of internal origin will not be identifiable, contrary to the external ones: daily variations and its main harmonics (12 hours, 8 hours and 6 hours), the 27-day variation due to the Sun's rotation and its main harmonics (27/2 days and 27/3 days), and the 1-year and 6-month periods and their harmonics associated to variations of the Sun's axis inclination relative to the Sun-Earth line. Some of these cycles are visible on Figure B.16. They contribute to the linear part of the time series' dynamics.

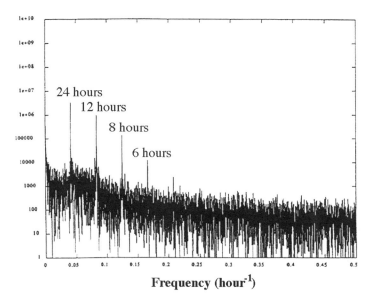

Figure B.16. Energy spectrum of the declination variations, for hourly measures made at the observatory of Chambon-la-Forêt (France) in 1988-1989.

The first stage consists in building the pseudo-phase space for this series. One possibility would of course be to use all the successive measures, but it was shown that a suitable sampling interval was needed to filter out most of the random noise associated to the measures. This is done using a *mutual information* criterion.

Mutual information is a measure of the general statistical dependency between the time series x(n) of possible values $S = \{ s_i \mid \ni n; (n) = s_i \}$ and the delayed time series x(n+T) of the values $\mathcal{D} = \{ d_i \mid \ni n; x(n+T) = d_i \}$.

The mutual information is defined by:

$$I(S,\mathcal{D}) = \mathcal{H}(S) + \mathcal{H}(\mathcal{D}) - \mathcal{H}(S,\mathcal{D}) \tag{B.49}$$

This is the sum of informations coming from the individual time series measured as entropy bits:

$$\begin{cases} \mathcal{H}(S) = - \sum_i P_S(s_i) \log_2 . P_S(s_i) \\ \mathcal{H}(\mathcal{D}) = - \sum_i P_\mathcal{D}(d_i) \log_2 . P_\mathcal{D}(d_i) \end{cases} \tag{B.50}$$

and reduced to the information obtained from the two sets:

$$\mathcal{H}(S,\mathcal{D}) = - \sum_i P_{S\mathcal{D}}(s_i,d_i) \log_2 . P_{S\mathcal{D}}(s_i,d_i) \tag{B.51}$$

Here, $P_S(s_i)$ and $P_\mathcal{D}(d_i)$ are the probabilities that the series x(n) and x(n+T) take values s_i and d_i, and $P_{S\mathcal{D}}(s_i,d_i)$ is the probability that the series x(n) and x(n+T) simultaneously take the values s_i and d_j.

One remarks that for independent series, $\mathcal{H}(S,\mathcal{D}) = \mathcal{H}(S) + \mathcal{H}(\mathcal{D})$ and therefore $I(S,\mathcal{D}) = 0$ and that for a null delay $I(S,S) = \mathcal{H}(S,S) = \mathcal{H}(S)$.

We use a recursive algorithm to estimate $I(S,Q)$ for different delays. The first local minimum of mutual information determines the value of T chosen as time delay (see Figure B.17). For the Chambon-la-Forêt time series, T = 7 hours.

The second stage of the analysis investigates the minimum dimension of the phase space (or dipping dimension) d_E.

Let us remember that:

$$\mathbf{x}(n) = [\, x(n), x(n+T), ..., x(n+d_E-1)/T \,] \tag{B.52}$$

To minimise this dimension, we study the behaviour of the orbit's close neighbours in the phase space when $d_E \rightarrow d_E +1$. If the number of false closest neighbours becomes null, one assumes the orbit did not fold in the space of dimension d_E. If the dimension is too small, the attractor's singularities induce a folding of the orbit.

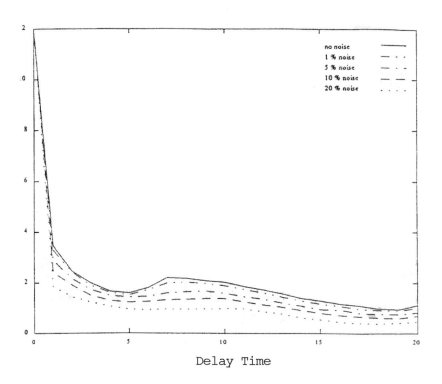

Figure B.17. Graphs showing the mutual information as a function of the time delay. The computations were performed for a Lorenz attractor (series of 20,000 values) with several levels of noise, for the Chambon-la-Forêt series of 14,500 values. From Hongre et al., 1994.

The square Euclidean distance for a closest neighbour $\mathbf{y}(n)$ of $\mathbf{x}(n)$ is:

$$R_d^2(n) = \sum_{k=0}^{d-1}(x(n+kT) - y(n+kT))^2 \tag{B.53}$$

Going to a dimension $(d+1)$ means adding a $(d+1)$-th coordinate to each vector $\mathbf{x}(n)$:

$$R_{d+1}^2(n) = R_d^2(n) + (x(n+dT) - y(n+dT))^2 \tag{B.54}$$

If the distance between $x(n)$ and $y(n)$ is large when going from dimension d to dimension (d+1) (i.e. larger than a threshold $R_{d+1}(k) - R_{d+1}(k)/R_d(k) > R_T$), $y(n)$ is said to be a "false closest neighbour". Practically, the threshold is chosen as $R_T \approx 10$.

The criterion can be increased again for noisy data with a finite length, if we call "false closest neighbours" points which are relatively close to a distance scale given by the attractor's size:

$$R_d(k)/\sigma_x \geq 2, \text{ where } \sigma_x = \left(\frac{1}{N-1} \sum_{n=1}^{N} (x(n) - \bar{x})^2 \right)^{1/2}$$

At this point in the analysis, we have reconstructed the phase space and estimated the attractor's topological invariants. We will now endeavour to estimate its metric invariants, i.e. its Lyapunov exponents, but in a situation where detailed information about the dynamics is lacking, as in a predictive model. We will use here a method proposed by Sano and Sawada (1985), looking at the evolution of small samples of sub-spaces with the tangent space along the orbit.

Let us consider a small perturbation $\delta x(0)$ of the orbit at time $t = t_0$ ($n = 0$). Its evolution follows:

$$x(n+1) + \delta x(n+1) = F(x(n) + \delta x(n)) \tag{B.55}$$

The following local approximation is valid:

$$\delta x(n+1) = DF(x(n)) \, \delta x(n) \tag{B.56}$$

The Jacobian $DF(x(n))$ is regarded here as a linear operator in the space tangent to $x(n)$.

This approximation can be rewritten as the combination of these operators along the orbit, to which was applied the initial perturbation:

$$\delta x(n+1) = DF(x(n)) \, DF(x(n-1)) \, ... \, DF(x(0)) \, \delta x(0) \tag{B.57}$$

$$= DF^{(n)}(x(0)) \, \delta x(n)$$

The ergodic multiplication theorem of Osceledec (1968) implies that, for an ergodic dynamic system with d dimensions, there is a limit operator $DF^{(n)}(x(0))$ independent of the initial choice for $x(0)$.

The spectrum of Lyapunov exponents is computed by approximating the tangent flow $DF^{(n)}(x(0))$ with a step-by-step map of a linear flow A_j, calculated with a least-mean squares method from a couple of "pseudo-tangent" vectors propagating in a given interval.

Using the orbit $x(j)$ as reference, the set of differences $y^i = x(j) - x(k_i)$, $i = 1, ..., N$ is called "pseudo-tangent". The N closest neighbours of $x(j)$ are inside a sphere of radius

ε_{max}: { $x(k_i)$ } = { $x(k_i)$; $|x(k_i) - x(j)| \leq \varepsilon_{max}$ }. Once an evolution interval $\Delta t = m\tau_S$ is selected, the differences y^i become the vectors $z^i = A_j y^i$.

The optimal estimation of the flow A_j is given by:

$$\min_{A_j} = \min_{A_j} \frac{1}{N} \sum_{i=1}^{N} |z^i - A_j y^i| \qquad (B.58)$$

Finally, the 1-D Lyapunov exponents are computed with:

$$\lambda_i = \lim_{n \to \infty} \frac{1}{n\tau_S} \sum_{j=1}^{N} \ln.|A_j e_i(j)| \qquad (B.59)$$

To test the method, it was applied to a Lorenz attractor on an orbit defined with 20,000 points. The results of this test are presented in Figure B.18. We know that the real values of Lyapunov exponents for the Lorenz attractor are (Wolf et al., 1984): $\lambda_1 = 1.5$; $\lambda_2 = 0.0 ...$; $\lambda_3 = -22.5$. There is a good agreement for the estimates of λ_1 and λ_2, but some imprecision on the highly negative value of λ_3. The free parameters used in this computation were: N = 20 for the number of closest neighbours in a sphere of radius $\varepsilon_{max} = 0.01$, m = 4 (0.08 s) for the evolution A_j, and a re-orthogonalisation interval of 0.3 s (every 15 points).

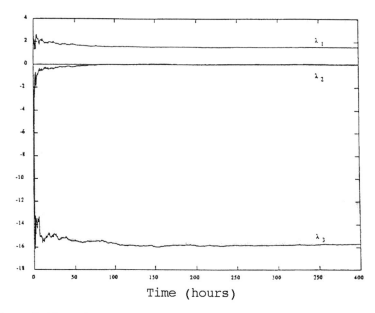

Figure B.18. Estimation of Lyapunov exponents for the Lorenz attractor, performed on a non-noisy orbit of 20,000 points. From Hongre et al., 1994.

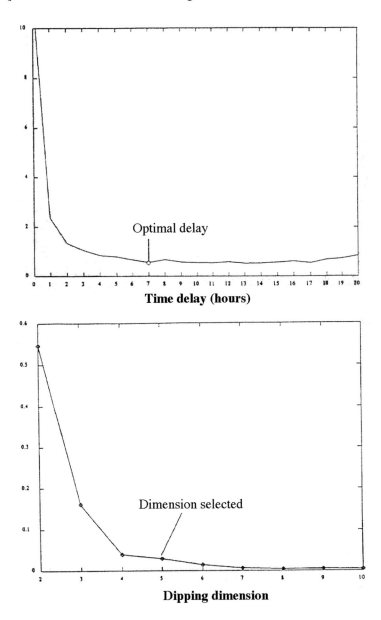

Figure B.19. The first graph shows the application of the mutual information criterion to the time series made from declinations recorded hourly at the Observatory of Chambon-la-Forêt, in 1988-89. The optimal time delay T is of 7 hours. The second graph shows the ratio of closest false neighbours, stabilising for a dipping dimension of 5.

The study of the time series made from the declinations recorded at the Observatory of Chambon-la-Forêt first focused on estimating the optimal time delay in the building of the phases space (Figure B.19). The mutual information criterion yielded a delay T = 7 hours. With the ratio of the closest false neighbours, stabilisation is observed to begin with a dipping dimension of 5.

This orbit in a 5-D space can be represented by being projected onto a hyperplan in 3 dimensions (Figure B.20).

The objective was the computation of the spectrum of Lyapunov exponents of dimension 1. Table B.3 and Figure B.21 sum up the results. The free parameters used in this study were N = 22, ε_{max} = 0.016, m = 7 hours, and a re-orthogonalisation interval of 11 hours.

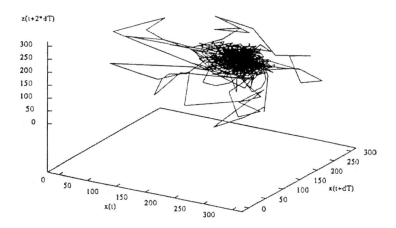

Figure B.20. Graphical representation of an attractor with a high dimension. The 5-D orbit in the phase space is projected onto a hyperplane in 3 dimensions.

The exponents have the respective signs of { +, 0, -, -, - }. There is at least one positive exponent, expressing the non-linear origin of this dynamics. The negative sum of exponents: $\sum_{i=1}^{5} \lambda_i$ = -0.147 confirms the dissipative nature of the source.

The inaccuracy on the double predictability time of the time series is: $\tau_{predict} \approx \ln. 2/\lambda_1$, and is estimated as: $\tau_{predict}$= 27 hours.

Table B.3. Values of Lyapunov exponents for the time series of Chambon-la-Forêt.

Exponent	Mean value	Standard deviation
λ_1	0.0259	0.186 10^{-2}
λ_2	0.0033	0.96 10^{-4}
λ_3	-0.0156	0.85 10^{-4}
λ_4	-0.0477	0.98 10^{-4}
λ_5	-0.1129	0.191 10^{-2}

Finally, we can verify the formula of Kaplan and Yorke, relating the information dimension of the time series and the values of the Lyapunov spectrum:

$$d_I = j + \left\{ \sum_{i=1}^{j} \lambda_i \,/\, |\lambda_{j+1}| \right\} \tag{B.60}$$

for which j satisfies the conditions:

$$\sum_{i=1}^{j} \lambda_i > 0 \text{ and } \sum_{i=1}^{j+1} \lambda_i < 0 \tag{B.61}$$

Let us remember that the series' information dimension is defined by:

$$d_I = \lim_{\varepsilon \to 0} \frac{I(\varepsilon)}{\log_2 \varepsilon} \tag{B.62}$$

where:

$$I(\varepsilon) = - \sum_{i=1}^{n(\varepsilon)} P_i \log_2 P_i \tag{B.63}$$

and P_i is the probability that the orbit passes through the i-th element of the set comprising $n(\varepsilon)$ to cover the whole orbit.

For the time series of Chambon-la-Forêt, computation shows that $d_I = 3.1$. Using the Lyapunov exponents, we find a value of $d_I = 3.12$, which is in good agreement.

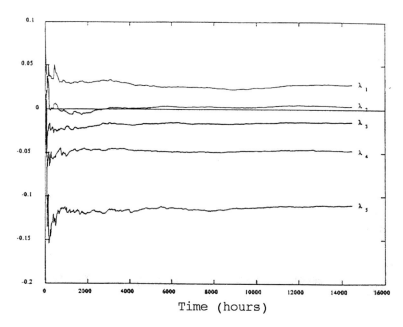

Figure B.21. Estimation of Lyapunov exponents, still for the time series of Chambon-la-Forêt (hourly measures).

B.5 PALAEONTOLOGICAL STUDIES OF EVOLUTION

The family of rodents named *Arcovidae* appeared 5 to 4.5 Ma ago. A systematic study allowed the gathering of information about 140 branches, a hundred of which still subsist (Chaline, 1987). Evolution scientists are wondering whether the distribution of speciation events (FAD: First Appearance Datum) is random, or it follows a power law.

This study was made possible by the existence of a relatively large series (around a hundred events), and the possibility of a precise dating of these events from their climatic/stratigraphic context (Chaline and Farjanel, 1990).

The method used was presented in Section 2.2.

Each event on the 1-D time scale is either the apparition (FAD) or extinction (LAD: Last Appearance Datum) of a particular species. The results from box-counting are represented in Figure B.22.

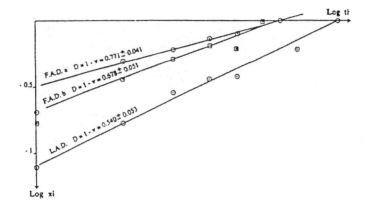

Figure B.22. Determinism in evolution. This graph shows that the apparitions and extinctions of rodent species during the Quaternary follow power laws. From Dubois et al., 1992.

These graphs show:

(1) a linear alignment of points in a time interval corresponding to approximately one order of magnitude.

(2) a significant difference between the slopes of regression lines, and therefore between the fractal dimensions of distribution laws:

$$D_{FAD} = 0.771 \pm 0.041 \text{ and } D_{LAD} = 0.540 \pm 0.053 \qquad (B.64)$$

This shows a greater clustering tendency for species extinctions than for the appearance of new species.

(3) The FAD process is self-similarity along time. The dimensions calculated for time periods of 1 Ma and 5 Ma are indeed very close.

One interpretation of these results is that the appearance of new species of *Arvicolidae* or their extinction is not random, and follows power laws. The differences between the dimensions D_{FAD} and D_{LAD} show that the greater clustering tendency of LADs can be interpreted as due to major external constraints, such as climatic variations (extinctions seem to originate from the warm or temperate phases following the major cooling episodes) (Chaline and Brunet-Lecomte, 1992).

This study brings back to mind the results obtained with the "game of life" and manifesting self-organised criticality (see Section 4.3).

Bibliography

> *Tempus erax rerum ...*
> *Time erases everything ...*
> Ovid, Metamorphoses

Aki, K.; "Re-evaluation of stress drop and seismic energy using a new model of earthquake faulting", in *Mécanismes et Prévision des Séismes*, C. Allègre (ed.), p. 23-50, CNRS: Paris, 1980

Aki, K.; "A probabilistic synthesis of precursory phenomena", in *Earthquake Prediction, an International Review*, D.W. Simpson and P.G. Richards (eds.), American Geophysical Union, Maurice Ewing Series, vol. 4, p. 566-574, 1981

Allan, D.W.; "On the behaviour of systems of coupled dynamos", Proc. Cambridge Phil. Soc., vol. 58, p. 671-693, 1962

Allen, C.R.; "Tectonic environments of seismically active and inactive areas along the San Andreas fault system", in *Proceedings of Conference on Geologic Problems of Sand Andreas Fault System*, W.R. Dickenson and A. Grantz (eds), p. 70-82, Stanford University, Stanford, Ca., 1968

Allègre, C.J., J.L. Le Mouël, A. Provost; "Scaling rules in rock fracture and possible implications for earthquake prediction", Nature, vol. 297, p. 47-49, 1982

Allègre, C.J., J.L. Le Mouël; "Introduction of scaling technics in brittle fracture of rocks", Physics of the Earth and Planetary Interiors, vol. 87, p. 85-93, 1994

Allègre, C.J., J.L. Le Mouël, Ha Duyen Chau, C. Narteau; "Scaling Organisation of Fracture Tectonics (SOFT) and earthquakes mechanism", Physics of the Earth and Planetary Interiors, vol. 92, p. 215-233, 1995

Argoul, F., A. Arnéodo, G. Grasseau, Y. Cagne, E.J. Hopfinger, U. Frisch; "Wavelet analysis of turbulence reveals the multifractal nature of the Richardson cascade", Nature, vol. 338, p. 51-53, 1989

Arnéodo, A., F. Argoul, G. Grasseau; "Transformation en ondelettes et renormalisation", in *Transformation en ondelettes, exposé*, vol. 9, p. 125-149, 1988

Arnéodo, A., G. Grasseau, M. Holschneider; "Wavelet transform of multifractals", Physics Review Letters, vol. 61, no. 20, p. 2281-2284, 1988

Arnéodo, A., D. Sornette; "Monte Carlo random walk experiments as a test of chaotic orbits of maps of the interval", Physics Review Letters, vol. 52, p. 1857-1869, 1984

Arnold, V.I.; "Catastrophe Theory", 3rd edition, Springer-Verlag: Berlin, 150 pp., 1992

Atten, P., J.G. Caputo; "Estimation expérimentale de dimension d'attracteurs et d'entropie", in *Traitement théorique des attracteurs étranges*, M. Cosnard (ed.), CNRS: Paris, p. 177-191, 1987

Aviles, C.A., C.H. Scholz, J. Boatwright; "Fractal analysis applied to characteristic segments of the San Andreas fault", Journal of Geophysical Research, vol. 92, p. 331-344, 1987

Avnir, D., D. Farin, P. Pfeifer; "Chemistry in non-integer dimensions between two and three, II : Fractal surface of adsorbents", J. Chem. Phys., vol. 79, p. 3566-3571, 1983

Backus, G.E., J.F. Gilbert; "Numerical application of a formalism for geophysical inverse problems", Geophysical Journal of the Royal Astronomical Society, vol. 13, p. 247-276, 1967

Bak, P., K. Chen; "The Physics of Fractals", Physica D, vol. 38, p. 5-12, 1989

Bak, P., K. Chen; "Self-Organized Criticality", Scientific American, vol. 54, p. 46-53, 1991

Bak, P., C. Tang; "Earthquakes as a self-organized critical phenomenon", Journal of Geophysical Research, vol. 94, p. 15,635-15,637, 1989

Bak, P., C. Tang, K. Weisenfeld; "Self-organized criticality: an explanation of 1/f noise", Physics Review Letters, vol. 59, p. 381-384, 1987

Bak, P., C. Tang, K. Weisenfeld; "Self-organized criticality in the Game of Life", Nature, vol. 342, p. 780-781, 1989

Ballu, V.; "Analyse fractale du fonctionnement de la dorsale médio-atlantique", M. Res. thesis, Univ. Paris-VII - IPGP, 1992

Barenblatt, G.I., A.V. Zhivago, P. Yu, P. Neprochnv, A. Ostrovskyl; "The fractal dimension: a quantitative characteristic of ocean bottom relief", Oceanology, vol. 24, p. 695-697, 1984

Barnsley, M.; "Fractals everywhere", Academic Press Inc.: San Diego, Ca., 394 pp., 1988

Barton, C.E; "Spectral analysis of palaeomagnetic time series and the geomagnetic spectrum", Philosophical Transactions of the Royal Society of London, vol. A-306, p. 203-209, 1982

Barton, C.E.; "Analysis of palaeomagnetic time series, techniques and applications", Geophysical Surveys, vol. 5, p. 335-368, 1983

Beauvais, A., J. Dubois, A. Badri; "Application d'une analyse fractale à l'étude morphométrique du tracé des cours d'eau: méthode de Richardson", C. R. Acad. Sci. Paris, vol. 318, no. II, p. 219-225, 1994

Bell, T.H.; "Statistical features of seafloor topography", Deep-Sea Research, vol. 22, p. 883-892, 1975

Bell, T.H.; "Mesoscale seafloor roughness", Deep-Sea Research, Part A, vol. 26, p. 65-76, 1978

Bennett, J.G.; "Broken coal", J. Inst. Fuel, vol. 10, p. 22-39, 1936

Bergé, P., Y. Pomeau, C. Vidal; "L'ordre dans le chaos: vers une approche déterministe de la turbulence", Hermann: Paris, 353 pp., 1984

Berry, M.V., J.H. Hannay; "Topography of random surfaces", Nature, vol. 273, p. 573, 1978

Bergstrom, B.H., D.D. Crabtree, C.L. Sollenberger; "Feed size effects in single particle crushing", Trans. Amer. Inst. Mining Metall. Petrol. Engrs., vol. 226, p. 433-441, 1963

Besicovitch, A.S.; "On the sum of digits of real numbers represented in the dyadic system (on sets of fractional dimensions)", Mathematische Annalen, vol. 110, p. 321-330, 1935

Besicovitch, A.S.; "On fundamental geometric properties of linearly measurable plane sets of points, III", Mathematische Annalen, vol. 116, p. 349-357, 1939

Besse, J. V. Courtillot; "Revised and synthetic apparent polar wander paths of the African, Eurasian, North American and Indian plates and true polar wander since 200 Ma", Journal of Geophysical Research, vol. 96, p. 4029-4050, 1991

Bond, F.C.; "The third theory of comminution", Trans. Amer. Inst. Mining Metall. Petrol. Engrs., vol. 193, p. 484-494, 1952

Bouligand, G.; "Ensembles impropres et nombre dimensionnel", Bull. Sci. Math. II, vol 52, p. 320-334 and p. 361-376, 1928

Bretherton, F.P., R.E. Davis, C.B. Flandry; "A technique for objective analysis and design of oceanographic experiments applied to MODE-73", Deep-Sea Research, vol. 23, p. 559-582, 1976

Broadbent, S.R., J.M. Hammersley; Proc. Camb. Phil. Soc., vol. 53, p. 629, 1957

Brown, S.R., C.H. Scholz; "Broad bandwidth study of the topography of natural rock surfaces", Journal of Geophysical Research, vol. 90, p. 12,575-12,582, 1985

Burridge, R., L. Knopoff; "Model and theoretical seismicity", Bull. Seism. Soc. Am., vol. 57, p. 341-371, 1967

Burrough, P.A.; "Fractal dimension of landscapes and other environmental data", Nature, vol. 294, p. 240-242, 1984

Camus, G., P.M. Vincent; "Krakatau Volcano", Sean Bull., vol. 10, 1981

Camus, G., A. Gourgaud, P.M. Vincent; "Petrologic evolution of Krakatau (Indonesia): implications for a future activity", J. Volcan. Geotherm. Res., vol. 33, p. 299-316, 1987

Cantor, G.; "Grundlagen einer allgemeinen Mannichfaltigkeitslehere", Mathematische Annalen, vol. 21, p. 545-591, 1883

Carta, S., R. Figari, G. Sartoris, E. Sassi, R. Scandone; "A statistical model for Vesuvius and its volcanological implications", Bull. Volcan., vol. 44, no. 2, p. 129-151, 1981

Casdagli, M.; "Non-linear production of chaotic time series", Physica D, vol. 35, p. 335-356, 1989

Chaline, J.; "Arvicolid data (*Arvicolidae, Rodentia*) and evolutionary concepts", Evolutionary Biology, vol. 21, p. 237-310, 1987

Chaline, J., P. Brunet-Lecomte; "Anatomie de la radiation européenne des Arvicolidés (*Rodentia*): un test quantifié du modèle des équilibres/désésquilibres ponctués", C. R. Acad. Sci. Paris II, vol. 314, p. 251-256, 1992

Chaline, J., G. Farjanel; "Plio-Pleistocene rodent biostratigraphy and palynology of Bresse Basin, France, and correlations within Western Europe", Boreas, vol. 19, p. 69-80, 1990

Charles, R.J.; "Energy-size reduction relationships in comminution", Trans. Amer. Inst. Mining Metall. Petrol. Engrs., vol. 208, p. 80-88, 1957

Charlier, C.V.L.; "Wie eine unendliche Welt ausgebaut sein kann", Arkiv für Mathematik, Astronomi och Fysik, vol. 16, p. 1-15, 1908

Cook, A.E.; "Two disc dynamos with viscous friction and time delay", Proc. Camb. Phil. Soc., vol. 71, p. 135-153, 1972

Courtillot, V., J.L. Le Mouël; "Time variations of the Earth's magnetic field: from daily to secular", Ann. Rev. Earth Planet. Sci., vol. 16, p. 389-476, 1988

Cox, A.; "Lengths of geomagnetic polarity intervals", Journal of Geophysical Research, vol. 73, p. 3247-3260, 1968

Cosnard, M.; "Traitement numérique des attracteurs étranges", CNRS: Paris, 276 pp., 1987

Croquette, V.; "Deux exemples mécaniques ayant des comportements chaotiques", in *Traitement numérique des attracteurs étranges*, M. Cosnard (ed.), CNRS: Paris, p. 107-127, 1987

Crutchfield, J.P., B.S. McNamara; "Equations of motion from a data series", Complex Systems, vol. 1, p. 417-452, 1987

Curry, J., J.A. Yorke; "A transition from Hopf bifurcation to chaos: computer experiments with maps in R2", in *The structure of attractors in dynamical systems*, Lecture Notes in Mathematics, vol. 668, p. 48, Springer-Verlag: Berlin, 1977

Davy, Ph., A. Sornette, D. Sornette; "Some consequences of a proposed fractal nature of continental faulting", Nature, vol. 348, p. 56-58, 1990

De Gennes, P.G.; J. Phys. Lett., vol. 37, no. L-1, 1976

Dehlinger, P.; "Marine Gravity", Elsevier: Amsterdam, 322 pp., 1978

Deutscher, G. R. Zallen, J. Adler; "Percolation structures and processes", Annals of the Israel Physical Society, vol. 5, Adam Hilger: Bristol, 1983

Ditto, W.L., S.N. Rauseo, M.L. Spano; "Experimental control of chaos", Phys. Rev. Lett., vol. 65, no. 26, p. 3211-3214, 1990

Donnison, J.R., R.A. Sugden; "The distribution of asteroid diameters", Mon. Not. R. Astron. Soc., vol. 210, p. 673-682, 1984

Dubois, J., J.L. Cheminée; "Application d'une analyse fractale à l'étude des cycles éruptifs du Piton de la Fournaise (La Réunion): modèle d'une poussière de Cantor", C. R. Acad. Sci. Paris II, vol. 307, p. 1723-1729, 1988

Dubois, J., L. Nouaili; "Quantification of the fracturing of the slab using a fractal approach", Earth and Planetary Science Letters, vol. 94, p. 97-108, 1989

Dubois, J., C. Pambrun; "Etude de la distribution des inversions du champ magnétique terrestre entre -165 Ma et l'actuel (échelle de Cox): Recherche d'un attracteur dans le système dynamique qui les génère", C. R. Acad. Sci. Paris II, vol. 311, p. 643-650, 1990

Dubois, J., J.L. Cheminée; "Fractal analysis of eruptive activity of some basaltic volcanoes", J. Volcan. Geotherm. Res., vol. 45, p. 197-208, 1991

Dubois, J., J. Chaline, P. Brunet-Lecomte; "Spéciation, extinction et attracteurs étranges", C.R. Acad. Sci. Paris II, vol. 315, p. 1827-1833, 1992

Dubois, J., C. Deplus, M. Diament; "Non-linear dynamics applied to Anak Krakatau eruptive activity: volcanological implications", J. Volcan. Geotherm. Res., vol. 47, 1992

Dubuc, B., J.F. Quiniou, C. Roques-Carmes, C. Tricot, S.W. Zucker; "Evaluating the fractal dimension of profiles", Phys. Rev. A, vol. 39, no. 3, p. 1500-1512, 1989

Eckman, J.P., S. Oliffson-Kamphorst, D. Ruelle, S. Ciliberto; "Lyapunov exponents from time series", Phys. Rev. A, vol. 34, no. 6, p. 4971-4979, 1986

Ershov, S.V., G.G. Malinetskii, A.A. Ruzmaikin; "A generalized two-disk dynamo model", Geophys. Astrophys. Fluid Dynamics, vol. 47, p. 251-277, 1989

Essam, J.W.; "Percolation Theory", Rep. Prog. Physics, vol. 43, p. 883-912, 1980

Evertsz, C.J.G., B. Mandelbrot; "Chaos and Fractals, Appendix B", Multifractal Measures, Springer-Verlag: Berlin, p. 921-953, 1992

Falconer, K.; "The geometry of fractal sets", Cambridge University Press: Cambridge, 1985

Falconer, K.; "The Hausdorff dimension of self-affine fractals", Math. Proc. Camb. Phil. Soc., vol. 103, p. 339-350, 1988

Falconer, K.; "Fractal geometry: mathematical foundations and applications", Wiley: Chichester, 288 pp., 1990

Farmer, J.D., E. Ott, J.A. Yorke; "The dimension of chaotic attractors", Physica D, vol. 7, p. 153-180, 1983

Farmer, J.D., D. Sidorowich; "Predicting chaotic time series", Phys. Rev. Lett., vol. 59, p. 845, 1987

Fatou, P.; "Sur les équations fonctionnelles", Bull. Soc. Math. France, vol. 47, p. 161-271 and vol. 48, p. 33-94 and 208-214, 1919

Fournier d'Albe, E.E.; "Two new worlds: I. The infra world; II. The supra world", Longmans Green: London, 1907

Fox, C.G., D.E. Hayes; "Quantitative methods for analysing the roughness of the seafloor", Rev. Geophys., vol. 23, p. 1-48, 1985

Filloux, J.H.; "Observation of very-low frequency electromagnetic signals in the ocean", J. Geomagn. Geoelectr., vol. 32, no. I, p. 1-12, 1980

Frisch, U., G. Parisi; in *Turbulence and predictability of geophysical fluid dynamics and climate dynamics*, Proc. Int. School of Phys. "Enrico-Fermi", M. Ghil: Amsterdam, 1985

Frohlich, C., R.J. Willemann; "Aftershocks of deep earthquakes do not occur preferentially on nodal planes of focal mechanisms", Nature, vol. 329, p. 41-42, 1987

Fujiwara, A., G. Kamimoto, A. Tsukamoto; "Destruction of basaltic bodies by high-velocity impact", Icarus, vol. 31, p. 277-288, 1977

Gabrielov, A.M. O.E. Dmitrieva, V.I. Keilis-Borok and co-workers; "Algorithms of long-term earthquake prediction", Int. School for Research Oriented to Earthquake-Prediction Algorithms, Software and Data Handling, p. 1-61, (available from CERESIS, Apt. 11363, Lima 14, Peru), 1986

Gallet, Y., J. Besse, L. Krystyn, J. Marcoux, H. Théveniaud; "Magnetostratigraphy of the late Triassic Bolucektasi Tepe section (South-Western Turkey): implications for changes in magnetic reversal frequency", Phys. Earth Planet. Inter., vol. 93, p. 273-282, 1992

Gaudin, A.M.; "An investigation of crushing phenomena", Trans. Amer. Inst. Mining Metall. Petrol. Engrs., vol. 73, p. 253-316, 1926

Geilikman, M.B., T.V. Golubeva, V.F. Pissarenko; "Multifractal patterns of seismicity", Earth Plan. Sci. Lett., vol. 99, p. 127-132, 1990

Gelfand, I.M., S.A. Guberman, V.I. Keilis-Borok, L. Knopoff, F. Press, E.Y. Ranzman, I.M. Rotwain, A.M. Sadovsky; "Pattern recognition applied to earthquake epicenters in California", Phys. Earth Planet. Inter., vol. 11., p. 227-283, 1976

Giardini, D., J.H. Woodhouse; "Horizontal shear flow in the mantle beneath the Tonga arc", Nature, vol. 307, p. 505-509, 1984

Gibert, D., V. Courtillot; "Seasat altimetry and the South-Atlantic geoid - Spectral Analysis", Journal of Geophysical Research, vol. 92, p. 6235-6248, 1987

Gilabert, A., M. Benayad, A. Sornette, D. Sornette, C. Vanneste; "Conductivity and rupture in crack-deteriorated systems", J. Phys. France, vol. 51, p. 247-257, 1990

Gilbert, L.E.; "Are topographic data sets fractal ?", Pure and Applied Geophysics, vol. 131, no. 1-2, p. 241-254, 1989

Gilbert, L.E., A. Malinverno; "A characterization of the spectral density of ocean floor topography", Geophys. Res. Lett., vol. 15, p. 1401-1404, 1988

Goslin, J., D. Gibert; "The geoid roughness: a scanner for isostatic process in oceanic areas", Geophysical Research Letters, vol. 17, no. 11, p. 1957-1960, 1990

Grassberger, P.; "Information flow and dimensions of strange attractors", in *Traitement Numérique des Attracteurs Etranges*, M. Cosnard (ed.), CNRS: Paris, p. 161-176, 1987

Grassberger, P., I. Procaccia; "Characterization of strange attractors", Phys. Rev. Lett., vol. 50, no. 5, p. 346-349, 1983

Grossman, A. J. Morlet; "Decomposition of Hardy functions into square integrable wavelets of constant shapes", SIAM J. Math. Anal., p. 723-736, 1984

Gresta, S., G. Patane; "Review of seismological studies at Mount Etna", Pure and Applied Geophysics, vol. 125, p. 951-970, 1987

Grunberger, D.; "Analyse fractale des réseaux de fractures simulés", M. Res. thesis, USTL-Montpellier, France, 18 pp., 1991

Guckenheimer, J.; Ann. Rev. Fluid Mech., vol. 18, p. 15-31, 1986

Gutenberg, B., C.F. Richter; "Seismicity of the Earth and associated phenomena", Princeton University Press: Princeton, NJ, 1949

Guyon, E., S. Roux, D.J. Bergman; "Critical behaviour of electric failure thresholds in percolation", Journal de Physique, vol. 48, p. 903-904, 1987

Gwinn, E.G. R.M. Westervelt; "Scaling structure of attractors at the transition from quasiperiodicity to chaos in electronic transport", Geophysics Review Letters, vol. 59, p. 157-160, 1987

Halsey, R.C., M.H. Jensen; "Spectra of scaling indices for fractal measures: Theory and experiment", Physica D, vol. 23, p. 112-117, 1986

Halsey, R.C., M.H. Jensen, L.P. Kadanoff, I. Procaccia, B.I. Shraiman; "Fractal measures and their singularities: The characterization of strange sets", Phys. Rev. A, vol. 33, p. 1141-1151, 1986

Hanks, T.C., H. Kanamori; "A moment magnitude scale", Journal of Geophysical Research, vol. 84, p. 2348-2350, 1979

Harris, C., R. Franssen, R. Loosveld; "Fractal analysis of fractures in rocks: the Cantor's dust method. - Comments", Tectonophysics, vol. 198, p. 107-115, 1991

Hartmann, W.K.; "Terrestrial, lunar and interplanetary rock fragmentation", Icarus, vol. 10, p. 201-213, 1969

Havstad, J.W., C.L. Ehlers; "Attractor dimension of nonstationary dynamical systems from small data sets", Phys. Rev. A, vol. 39, no. 2, p. 845-853, 1989

Hausdorff, F.; "Dimension und ausseres Mass", Mathematische Annalen, vol. 79, p. 157-179, 1919

Hawkins, G.S.; "Asteroidal fragments", Astrophysical Journal, vol. 65, p. 318-322, 1960

Hentschel, H.G.E., I. Procaccia; "The infinite number of generalized dimensions of fractals and strange attractors", Physica D, p. 435-444, 1983

Held, G.A., D.H. Solina, D.T. Keane, W.G. Haag, P.M. Horn, G. Grinstein; "Experimental study of critical-mass fluctuations in an evolving sandpile", Phys. Rev. Letters, vol. 65, no. 9, p. 1120-1123, 1990

Hénon, M.; "A two-dimensional mapping with a strange attractor", Communications in Mathematics and Physics, vol. 50, p. 69, 1976

Hongre, L., M. Zhizhin, J. Dubois; "Chaotic characteristics of the magnetic field: fractal dimensions and Lyapunov exponents", Cahiers du Centre Européen de Géodynamique et de Sismologie, 1994

Houry, S., J.F. Minster, C. Brossier, K. Dominh, C. Gennero, A. Cazenave, P. Vincent; "Radial orbit error reduction and mean sea surface computations from the Geosat altimeter data", Journal of Geophysical Research, vol. 99, p. 4519-4531, 1994

Huang, J., D.L. Turcotte; "Fractal mapping of digitized images: application to the topography of Arizona and comparisons with synthetic images", Journal of Geophysical Research, vol. 94, p. 7491-7495, 1989

Hulot, G.; "Observations géomagnétiques et géodynamo", Ph.D. Thesis, Univ. Paris-VII, 377 pp., 1992

Hulot, G., J.L. Le Mouël, J. Wahr; "Taking into account truncation problems and geomagnetic model accuracy in assessing computed flows at the core-mantle boundary", Geophysical Journal International, vol. 108, p. 224-246, 1992

Hwa, T., M. Kardar; "Dissipative transport in open system: an investigation of Self-Organized Criticality", Phys. Rev. Lett., vol. 62, no. 16, p. 1813-1816, 1989

Ishimoto, M., K. Iida; "Observations sur les séismes enregistrés par le microséismographe construit dernièrement", I. Bull. Earthqu. Res. Inst., vol. 17, p. 443-478, 1939

Jensen, M.H., L.P. Kadanoff, A. Libchaber, I. Procaccia, J. Stavans; "Global universality at the onset of chaos: results of a forced Rayleigh-Bénard experiment", Phys. Rev. Lett., vol. 55, p. 2798-2801, 1985

Julia, G.; "Mémoire sur l'itération des fonctions rationnelles", Journal de Mathématiques Pures et Appliquées, vol. 4, p. 47-245, 1918

Kahane, J.P.; "The technique of using random measures and random sets in harmonic analysis", in *Advances in Probability and Related Topics*, P. Ney (ed.), vol. 1, p. 65-101, Marcel Dekker: New York, 1971

Kahane, J.P.; "Measures and dimensions. Turbulences and Navier-Stokes equations", Lecture Notes in Mathematics, vol. 565, p. 94-103, Springer: New-York, 1976

Kanamori, H., D.L. Anderson; "Theoretical basis of some empirical relations in seismology", Bulletin of the Seismological Society of America, vol. 65, p. 1073-1096, 1975

Kaplan, J., J. Yorke; "Functional differential equations and the approximation of fixed points", Proceedings, Bonn - July 1978, Lecture Notes in Mathematics, vol. 730, p. 228, Springer: Berlin, 1978

Kaufman, R.; "On the Hausdorff dimension of projections", Mathematika, vol. 15, p. 153-155, 1968

Keilis-Borok, V.I.; "The lithosphere of the Earth as a non-linear system with implications for earthquake prediction", Rev. Geophys., vol. 28, no. 1, p. 19-34, 1990

Keilis-Borok, V.I., V.G. Kossobokov; "A complex of long-term precursors for strongest earthquakes in the world", Proceedings of the 27th Geological Congress: Earthquakes and hazard prevention, vol. 64, p. 56-61, Nauka: Moscow, 1984

Keilis-Borok, V.I., V.G. Kossobokov; "Time of increased probability for the great earthquakes of the world", Comput. Seismol., vol. 19, p. 48-58, 1986

Keilis-Borok, V.I., V.G. Kossobokov; "Premonitory activation of seismic flow: Algorithm M8", Lecutre Notes of the Workshop on Global Geophysical Informations with Applications to Earthquake Prediction and Seismic Risk, Rep. H4, SMR/303-10, International Centre for Theoretical Physics, Trieste, Italy, 17 pp., 1988

Keilis-Borok, V.I., V.G. Kossobokov; "Times of increased probability of strong earthquakes ($M > 7.5$), diagnosed by algorithm M8 in Japan and adjacent territories", Journal of Geophysical Research, vol. 95, p. 12,413-12,422, 1990

Keilis-Borok, V.I., I.M. Rotwain; "Diagnosis of time of increased probability of strong earthquakes in different regions of the world: Algorithm CN", Phys. Earth Planet. Int., vol. 61, p. 57-72, 1990

Keilis-Borok, V.I., L. Knopoff, I.M. Rotwain, C.R. Allen; "Intermediate-term prediction of times of occurrence of strong earthquakes in California and Nevada", Nature, vol. 335, p. 690-694, 1988

Kerr, R.A.; "Perestroika comes to earthquake forecasts", Science, vol. 251, p. 1314-1316, 1991

Klein, F.W.; "Patterns of historical eruptions at Hawaiian volcanoes", Journal of Volcanology and Geothermal Research, vol. 12, p. 1-35, 1982

Knuth, D.E.; "Fundamental algorithms, the art of computer programming", Addison-Wesley: Reading, Mass., 634 pp., 1973

Kolmogorov, A.N.; "Local structure of turbulence in an incompressible liquid for very large Reynolds numbers", C. R. Acad. Sci. USSR, vol. 30, p. 299-303, 1941

Kono, M.; "Geomagnetic polarity changes and the duration of volcanism in sucessive lava flows", Journal of Geophysical Research, vol. 78, p. 5972-5982, 1973

Korcak, J; "Deux types fondamentaux de distribution statistique", Bull. Inst. Intern. Stat., vol. III, p. 295-299, 1938

Kuhn, T.; "La structure des révolutions scientifiques", Flammarion: Paris, 286 pp., 1970

Landau, L., E. Lifschitz; "Physique Statistique", Mir: Moscow, 583 pp., 1967

Landau, L., E. Lifschitz; "Mécanique", Mir: Moscow, 228 pp., 1969

Landau, L., E. Lifschitz; "Mécanique des fluides", Mir: Moscow, 669 pp., 1971

Lange, M.A., T.J. Ahrens, M.B. Boslough; "Impact cratering and spall failure of gabbro", Icarus, vol. 58, p. 383-395, 1984

Lautman, A.; "Essai sur l'unité des mathématiques et divers écrits", Union Générale d'Editions: Paris, 319 pp., 1977

Lebesgue, H.; "Remarques sur les théories de la mesure et de l'intégration", Annales de l'Ecole Normale Supérieure, vol. 35, p. 191-250, 1918

Ledésert, B. J. Dubois, B. Velde, A. Meunier, A. Genter, A. Badri; "Geometrical and fractal analysis of a three-dimensional hydrothermal vein network in fractured granite", Journal of Volcanology and Geothermal Research, vol. 56, p. 267-280, 1993a

Ledésert, B., J. Dubois, A. Genter, A. Meunier; "Fractal analysis of fractures applied to Soultz-sous-Forêts hot dry rock geothermal prgram", Journal of Volcanology and Geothermal Research, vol. 57, p. 1-17, 1993b

Lévy, P.; "Sur la possibilité d'un univers de masse infinie", Ann. Phys., vol. 14, p. 184-189, 1930

Liebovitch, L.S., T. Toth; "A fast algorithm to determine fractal dimensions by box counting", Phys. Lett. A, vol. 141, p. 386-390, 1989

Lorenz, E.N.; "Deterministic non-periodic flow", Journal of Atmospheric Sciences, vol. 20, p. 130, 1963

Malinverno, A., L.E. Gilbert; "A stochastic model for the creation of abyssal hill topography at a slow spreading center", Journal of Geophysical Research, vol. 94, p. 1665-1675, 1989

Malinverno, A., P.A. Cowie; "Normal faulting and topographic roughness of mid-ocean ridge flanks", Journal of Geophysical Research, vol. 98, p. 17,921-17,939, 1993

McCrosky, R.E.; "Distribution of large meteoric bodies", Spec. Rep. 280, Smithson. Astrophys. Observa., Washington DC, 1968

MacKey, M.C., L. Glass; "Oscillation and chaos in physiological control systems", Science, vol. 197, p. 287-289, 1977

Madden, T.R.; Geophysics, vol. 41, p. 1104, 1976

Madden, T.R.; "Microcrack connectivity in rocks: a renormalization group approach to the critical phenomena of conduction and failure in crystalline rocks", Journal of Geophysical Research, vol. 88, p. 585-592, 1983

Main, I.G.; "A characteristic earthquake model of the seismicity preceding the eruption of Mount St. Helens on 18 May 1980", Phys. Earth Planet. Int., vol. 49, p. 283-293, 1987

Malraison, B., P. Atten, P. Bergé, M. Dubois; "Dimension d'attracteurs étranges: une détermination expérimentale en régime chaotique de deux systèmes convectifs", C. R. Acad. Sci. Paris, vol. 297, p. 209-214, 1983

Mandelbrot, B.B.; "Some noises with 1/f spectrum, a bridge between direct current and white noise", IEEE Transactions on Information Theory, vol. 13, p. 289-298, 1967

Mandelbrot, B.B.; "How long is the coast of Britain? Statistical self-similarity and fractional dimension", Science, vol. 155, p. 636-638, 1967

Mandelbrot, B.B.; "Les objets fractals", Flammarion: Paris, 203 pp., 1975

Mandelbrot, B.B.; "Les images fractales: un art pour l'amour de la science et ses applications", Sciences et Techniques, no. 16-19, p. 34-35, 1984

Mandelbrot, B.B.; "Multifractal measures, especially for the geophysicist", Pure and Applied Geophysics, vol. 131, p. 5-42, 1989

Manneville, P.; "Structures dissipatives, chaos et turbulence", Collection Aléa Saclay, Commissariat à l'Energie Atomique: Gif-sur-Yvette, France, 417 pp., 1991

Marstrand, J.M.; "Some fundamental geometrical properties of plane sets of fractional dimensions", Proc. London Math. Soc., vol. 3, no. 4, p. 257-302, 1954

Maruyama, T.; "Frequency distribution of the sizes of fractures generated in the branching process: elementary analysis", Bull. Earthqu. Res. Inst., vol. 53, p. 407-421, 1978

Mathis, J.S.; "The size distribution of interstellar particles, II: Polarization", Astrophysical Journal, vol. 232, p. 747-753, 1979

Mattila, P.; "Hausdorff dimension, orthogonal projections and intersections with planes", Ann. Acad. Sci. Fennicae, vol. A1, p. 227-244, 1975

Mattila, P.; "Hausdorff dimension and capacities of intersections of sets in n-space", Acta Mathematicae, vol. 152, p. 77-105, 1984

Mattila, P.; "On the Hausdorff dimension and capacities of intersections", Mathematika, vol. 32, p. 213-217, 1985

McFadden, P.L.; "Statistical tooks for the analysis of geomagnetic reversal analysis", Journal of Geophysical Research, vol. 89, p. 3363-3372, 1984

McFadden, P.L., R.T. Merrill; "Lower mantle convection and geomagnetism", Journal of Geophysical Research, vol. 89, p. 3354-3362, 1984

Meakin, P., H.E. Stanley, A. Coniglio, T.A. Witten; "Surfaces, interfaces and screening of fractal structures", Phys. Rev. A, vol. 32, p. 2364-2369, 1985

Menke, W.; "Geophysical data analysis: Discrete inverse theory", Academic Press: San Diego, Ca., 289 pp., 1989

Minkowski, H.; "Über die Begriffe Lange, Ooberflache und Volumen", Jahresbericht der Deutschen Mathematikvereinigung, vol. 9, p. 115-121, 1901

Mogi, K.; "Study of the elastic shocks caused by the fracture of heterogeneous materials and its relation to earthquake phenomena", Bull. Earthqu. Res. Inst., vol. 40, p. 125-173, 1962

Moller, M., W. Lange, F. Mitschke, N.B. Abraham, U. Hubner; "Errors from digitizing and noise in estimating attractor dimensions", Phys. Lett. A, vol. 138, p. 176-182, 1989

Nagahama, H., K. Yoshii; "Scaling laws of fragmentation", in *Fractals and Dynamic Systems in Geosciences*, p. 25-36, J. Kruhl (ed.), Springer-Verlag: Berlin, 421 pp., 1994

Normand, J.M., H.J. Hermann, M. Hajjar; J. Stat. Phys., vol. 52, p. 747, 1988

Norton, V.A.; "Generation and display of geometrical fractals in 3-D", Computer Graphics, vol. 16, p. 61-67, 1982

Nouaili, L., J. Dubois, C. Deplus; "Analyse fractale de la séismicité de la zone de subduction des îles Tonga", C. R. Acad. Sci. Paris II, vol. 305, p. 1357-1364, 1987

Ohnaka, M., K. Mogi; "Frequency dependence of acoustic emission activity in rocks under incremental, uniaxial compression", Bull. Earthqu. Res. Inst., vol. 56, p. 67-89, 1981

Okubo, P.G., K. Aki; "Fractal geometry in the San Andreas fault system", Journal of Geophysical Research, vol. 92, p. 345-355, 1987

Oseledets, V.; "Multiplicative ergodic theorem. Lyapunov characteristic numbers for dynamic systems", Transactions of the Moscow Mathematical Society, vol. 19, p. 197-231, 1968

Otsuka, M.; "A chain-reaction type source model as a tool to interpret the magnitude-frequency relation of earthquakes", J. Phys. Earth, vol. 20, p. 35-45, 1972

Ott, E., C. Grebogi, J.A. York; "Controlling chaos", Phys. Rev. Lett., vol. 64, no. 11, p. 1196-1199, 1990

Panteleyev, A.N., J. Dubois, M. Diament; "Roughness of the global altimetric geoid", Cahiers du Centre Européen de Géodynamique et de Sismologie, 1994

Panteleyev, A.N., J. Dubois; "A scale-invariant analysis of the global altimetric geoid", Nonlinear Processes in Geophysics, vol. 5, p. 43-50, 1994

Parsons, B., J.G. Sclater; "An analysis of the variation of ocean floor bathymetry and heat flow with age", Journal of Geophysical Research, vol. 82, p. 803-827, 1977

Pecora, L.M., T.L. Carroll; "Synchronization in chaotic systems", Phys. Rev. Lett., vol. 64, no. 8, p. 821-824, 1990

Pisarenko, V.F., D.V. Pisarenko; Phys. Lett. A, vol. 153, p. 169, 1991

Pisarenko, D.V., V.F. Pisarenko; "Statistical estimation of the correlation dimension", Phys. Lett. A, vol. 197, p. 31-39, 1995

Pontrjagin, L., L. Schnirelman; "Sur une propriété métrique de la dimension", Annals of Mathematics, vol. 33, p. 156-162, 1932

Poincaré, H.; "Méthodes nouvelles de la mécanique céleste, 1(1892)", Dover: New York, 1957

Poincaré, H.; "Science et méthode", 1st ed., Flamarion: Paris, 1909

Prigogine, I.; "Les lois du chaos", Nouvelle Bibliothèque des Sciences, Flammarion: Paris, 127 pp., 1993

Prigogine, I., T.Y. Petrovsky, H.H. Hasegawa, S. Tasaki; "Integrability and chaos in classical and quantum mechanics", Chaos, Solitons and Fractals, vol. 1, no. 1, p. 3-24, 1991

Ramsey, J.B., H.J. Yuan; "The statistical properties of dimension calculations using small data sets:, Nonlinearity, vol. 3, p. 155-176, 1990

Roberts, R.G.; "The deep electrical structure of the Earth", Geophysical Journal of the Royal Astronomical Society, vol. 85, p. 583-600, 1986a

Roberts, R.G.; "Global electromagnetic induction", Geophysical Surveys, vol. 8, p. 339-374, 1986b

Roux, S.; "Structures et désordres", Ph.D. thesis, Ecole Nationale des Ponts et Chaussées: Paris, 510 pp., 1990

Richardson, L.F.; "The problem of contiguity: an appendix of statistics of deadly quarrels", General Systems Yearbook, vol. 6, p. 139-187, 1961

Rikitake, T.; "Oscillations of a system of disk dynamos", Proc. Camb. Phil. Soc., vol. 54, p. 89-105, 1958

Ruelle, D., F. Takens; "On the nature of turbulence", Com. Mathem. Phys., vol. 20, p. 167-192 and vol. 23, p. 343-344, 1971

Sayles, R.S., T.R. Thomas; "Surface topography as a non-stationary random process", Nature, vol. 271, p. 431-434, 1978

Scholz, C.H.; "The frequency- magnitude relation of microfracturing in rock and its relation to earthquakes:, Bulletin of the Seismological Society of America, vol. 58, p. 399-416, 1968

Schopenhauer, A.; "Le monde comme volonté et comme représentation (1844)", PUF: Paris, 1,434 pp., 1966

Schuhmann, R., Jr.; "Energy input and size distribution in comminution", Trans. Amer. Inst. Mining Metall. Petrol. Engrs., vol. 217, p. 22-25, 1960

Shaw, H.R.; "Uniqueness of volcanic systems", in *Volcanism in Hawaii*, US Geological Survey Professional Papers, p. 1357-1394, 1988

Shaw, H.R., B. Chouet; "Singularity spectrum of intermittent seismic tremor at Kilauea volcano, Hawaii", Geophysical Research Letters, vol. 16, no. 2, p. 195-198, 1989

Schoutens, J.E.; "Empirical analysis of nuclear and high-explosive cratering and ejecta", Nuclear Geophysics Sourcebook, vol. 55, part 2, section 4, Rep. DNA 65/01H-4-2, Defence Nuclear Agency: Bethesda, Md., 1979

Shebalin, P., N. Girardin, I. Rotwain, V. Keilis-Borok, J. Dubois; "Local inversion of active and non-active seismic zones before earthquakes with magnitude M ≥ 5 in Lesser Antillean arc", Physics of the Earth and Planetary Interiors, vol. 97, p. 163-175, 1995

Shinbrot, T., C. Grebogi, E. Ott, J.A. Yorke; "Using small perturbations to control chaos", Nature, vol. 363, p. 411-417, 1993

Smalley, R.F., D.L. Turcotte, S.A. Solla; "A renormalization group approach to the stick-slip behavior of faults", Journal of Geophysical Research, vol. 90, p. 1894-1900, 1985

Smalley, R.F., J.L. Chatelain, D.L. Turcotte, R. Prévot; "A fractal approach to the clustering of earthquakes: applications to the seismicity of the New Hebrides", Bulletin of the Seismological Society of America, vol. 77, no. 4, p. 1368-1381, 1987

Smith, D.K., T.H. Jordan; "Seamount statistics in the Pacific Ocean", Journal of Geophysical Research, vol. 93, p. 2899-2917, 1988

Sornette, A.; "Lois d'échelle dans les milieux fissurés, application à la lithosphère", Ph.D. thesis, Univ. Paris-XI, 250 pp., 1990

Sornette, A., D. Sornette; "Self-Organized Criticality and earthquakes", Europhys. Lett., vol. 9, p. 197-202, 1989

Sornette, D., P. Davy, A. Sornette; "Structuration of the lithosphere in plate tectonics as a Self-Organized Criticality phenomenon", Journal of Geophysical Research, vol. 95, p. 17,353-17,361, 1990

Sornette, A., J. Dubois, J.L. Cheminée, D. Sornette; "Are sequences of volcanic eruptions deterministically chaotic?", Journal of Geophysical Research, vol. 96, p. 11,931-11,945, 1991

Sugihara, G., R.M. May; "Non-linear forecasting as a way of distinguishing chaos from measurement error in time series", Nature, vol. 344, p. 734-741, 1990

Tait, S., C. Jaupart, S. Vergniolle; "Pressure, gas content and eruption periodicity of a shallow crystallising magma chamber", Earth and Planetary Science Letters, vol. 92, p. 107-123, 1989

Takens, F.; "Dynamical systems and turbulence", Warwick Lecture Notes in Math., vol. 898, p. 366, Springer: Berlin, 1980

Takens, F.; "On the numerical determination of the dimension of a strange attractor", Lecture Notes in Math., vol. 1125, p. 110-128, Springer: Berlin, 1985

Tarantola, A., B. Valette; "Generalized non-linear inverse problems solved using the least squares criterion", Rev. Geophys. Space Phys., vol. 20, p. 219-232, 1982

Tartaron, F.X.; "A general theory of comminution", Trans. Amer. Inst. Mining Metall. Petrol. Engrs., vol. 226, p. 183-190, 1963

Thom, R.; "Stabilité structurelle et morphogénèse: essai d'une théorie générale des modèles", Inter-Editions: Paris, 351 pp., 1972

Thom, R.; "Paraboles et catastrophes", Flammarion: Paris, 189 pp., 1989

Thouveny, N.; "Variations du champ magnétique terrestre au cours du dernier cycle climatique (depuis 120 000 ans)", CERLAT Memoirs, no. 3, 349 pp., 1991

Turcotte, D.L.; "Fractals and fragmentation", Journal of Geophysical Research, vol. 91, p. 1921-1926, 1986

Turcotte, D.L.; "Fractals in Geology and Geophysics", Pure and Applied Geophysics, vol. 131, p. 171-196, 1989

Turcotte, D.L.; "Fractals and Chaos in Geology and Geophysics", Cambridge University Press: New York, 221 pp., 1992

Valéry, P.; "Oeuvres", NRF-La Pléiade: Paris, 1,829 pp., 1966

Velde, B., J. Dubois; "Fractal analysis of fractures in rocks: the Cantor's dust method - Reply", Tectonophysics, vol. 198, p. 112-115, 1991

Velde, B., J. Dubois, G. Touchard, A. Badri; "Fractal analysis of fractures in rocks: the Cantor's dust method", Tectonophysics, vol. 179, p. 345-352, 1990

Velde, B., J. Dubois, D. Moore, G. Touchard; "Fractal patterns of fractures in granites", Earth and Planetary Science Letters, vol. 104, p. 25-35, 1991

Vere-Jones, D.; "A branching model of crack propagation", Pure and Applied Geophysics, vol. 114, p. 711-725, 1976

Voss, R.F.; "1/f noise in music; music from 1/f noise", Journal of the Acoustical Society of America, vol. 63, p. 258-263, 1978

Voss, R.F.; "Random fractal forgeries", in *Fundamental Algorithms for Computer Graphics*, R.A. Earnshaw (ed.), NATO ASI F13, vol. 13-16, p. 805-835, Springer: New York, 1985

Wolf, A. J.B. Swift, H.L. Swinney, J.A. Vastano; "Determining Lyapunov exponents from a time series", Physica D, vol. 16, p. 285-317, 1985

Wickman, F.E.; "Repose period pattern of volcanoes", Arkiv Mineral. Geol., vol. 4, no. 7, p. 291-364, 1966

Wickman, F.E.; "Markov models of repose-period patterns of volcanoes", in *Random Processes in Geology*, p. 135-161, Springer Verlag: Berlin, 1976

Wittgenstein, L.; "Tractatus logico-philosophus (1918)", P. Klossowski (translation), Gallimard: Paris, 364 pp., 1961

Index

affine   83, 84
affine (self-)   83
algorithm
    Barnsley determinist   161, 162
    Barnsley random   161, 162, 163
    CN   7, 129, 133-136, 219-223
    M8   7, 129, 130-133, 219-223
altimetry   81
anisotropy (fracturation)   99
asthenosphere   111
Atten and Caputo (relation)   43, 146
attraction (basin)   38
attraction (process)   37, 38, 40
attractor
    aperiodic   39, 40
    Feigenbaum   48
    Hénon   45, 48
    Lorenz   44, 45
    Mackey-Glass   48, 49
    Rössler   47, 48

       strange   3, 7, 39-41, 43-49, 78, 79
avalanche   126, 128
bifurcation (cascade)   58
blob   217, 218
Borelian   9, 12, 28
box-counting   22, 25, 26, 28, 97, 101, 197, 198, 232
butterfly effect   183
Cantor (target)   29
Cantor (dust)   7, 16, 19, 24, 60, 97, 98, 101, 111, 116, 119-121
chaos   2, 31, 37, 41, 55-57, 71-80, 119, 123, 129, 139, 142, 150-153, 157, 181-187
cluster   98, 103-105, 116, 119, 210-211, 214-217
cluster (infinite)   216-218
clustering   6, 106, 116-118, 130, 131, 133, 134, 150, 154, 172, 220, 233
coast length   5, 15, 82
conductivity   104
contraction (area)   38, 39, 55
correlation function   42, 44, 120-122
covering   11, 17-103
Cox scale   138, 140-143
critic state   126-129
       self-organised   126-129
critical exponents
       percolation   214-216
       Lyapunov   3, 46, 49-53, 139, 227, 231
degree of freedom   7, 31-37, 181-182
density (local)   178, 179
density (spectral energy)   85
dilation   37, 38
dimension
       generalised   2, 59-62
       correlation   2, 60
       fractal   2, 17, 41, 98, 99, 110, 111, 174, 196-198, 202
       information   2, 60
       Hausdorff   2, 13-17, 173-175

Index

    Hausdorff-Besicovitch   2, 13, 15
    Lebesgue   2
    Lyapunov   52
    of a fractal set   17
    dipping   229
    roughness   114-116
    similarity   2, 59, 60
    Euclidean   14, 225
    structural   197
    textural   197
    notion   4, 13-17
    Rényi   2, 6-62
distribution   88
    cumulated   88, 107-110
    exponential   88, 133
    Gamma   133, 140
    normal   107
    Poisson   107, 116, 133
    Weibull   88, 95
dominoes (game of)   126-128, 171
dynamo (theory)   137
eigenvalues   72
entropy   60, 225
evolution (theory of)   232
fault   96, 101, 208
fern leaf   164
fisssuration   6, 103
Floquet (matrix)   72-74
flow   7, 54, 228
    seismic, eruptive   7, 129
force field (central and non-central)   34-37
forecasting   222
    short-term   152-153
    earthquakes   219-223

256   Index

        volcanic eruptions   119-125
fractal (analysis)   5, 116, 195-197
fractal (geometry)   6, 87, 103
fractal (object)   1, 109
fractal (set)   3, 7, 11, 15-31, 159-169
        Borelians   10, 179
        Cantor   17-21, 98, 109
        random Cantor   21, 22
        Fatou   166
        Feigenbaum   48
        Hénon   45-47, 184-185
        Julia   165-167
        Mackey-Glass   48, 49
        Mandelbrot   167, 168, 175-177
        Rössler   47, 48
        definition   17
        regular, irregular   180, 181
        triadic Cantor   17-21, 109
fractal (tree)   165
fracturation   4, 30, 81, 87, 95-101, 113, 149, 197, 203-210
        rock fracturation   4, 81, 87
fractures   5, 28, 95-101, 202, 203, 205-210
fragmentation   3, 4, 6, 81, 87-95, 113-116, 149
game
        dominoes   126-128, 171
        "game of life"   171-172, 233
Gaussian   68
geoid
        altimetric   87
        roughness   86, 198-202
geomagnetic field intensity   137-143
geomagnetic field inversions   137-143
geomagnetic field time variations   137-148
geomagnetism   137-148, 224-231

geomorphology   82-86, 195-198
Gutenberg and Richter law   5, 107-112, 114, 115
Hausdorff
    measure   2, 11, 12, 173, 174
    dimension   2, 13-15, 174, 175
histogram   107-112
Hölder (condition)   175, 176
Hölder (function)   175, 176
Hopf (bifurcation)   74-76
hydrographic network   195-197
infimum   11
information (mutual)   225-228
intermissions   6, 116-118
intersection (of fractal sets)   29, 30, 100
invariants (scale)   109, 128, 129
IFS (Iterated Function System)   160-165
IFS code
    fern leaf   164
    fractal tree   165
    ivy leaf   162
    Sierpinski's triangle   163
    Sierpinski's square   164
iterations   53, 92, 60-165, 170, 187, 204
ivy (leaf)   162
Jacobian   46, 49, 50, 227
Kaplan and Yorke formula   139
Kolmogorov-Smirnov test   118
law
    Gutenberg and Richter   5, 107-112, 114, 115
    Poisson   107-113, 116
    power   7, 86, 107-113, 116, 233
    Walker-Lewis   114
Legendre polynomials   147
Lie derivative   46

limit cycle   37-58, 71, 139
Lipschitz   175
lithosphere   97, 111, 113, 114, 126, 128, 129
local density   178, 179
log-log graph   16, 42, 87, 98, 99
Lyapunov
    exponent   3, 46, 49-53, 139, 227, 231
    dimension   52
map
    map (first-return)   4, 6, 53-55, 122-124
    map (Ikeda)   153
    map (Lipschitz)   175, 176
    map (logistic)   55-58
    map (quadratic)   57
Markovian (chain)   118
Menger (sponge)   23, 25
measure
    Hausdorff   2, 11, 12, 173, 174
    Lebesgue   2, 10-11, 174
    multifractal   3, 52-71, 188-192
    Richardson   195
method
    Minkovski-Bouligand   198
    Grassberger and Proccacia   7, 42-44, 119, 141, 142, 145
microscope (mathematical)   68
mid-ocean ridge   82, 193
models
    Allègre and LeMouël   203-208
    Allègre et al. (SOFT)   209
    branching   111
    Burridge and Knopoff   6
    Curry and Yorke   77-79
    Ershov et al.   7, 138-139
    fractionary Brownian   4

Index

    sand heap   126-129, 171
    SOFT   209
    "sound pillar"   91-93
    weakness plane   93, 94
    Rikitake   7, 138
Moebius strip   48
moment
    kinetic   34, 35
    seismic   109
    statistical   189-192
movement
    Brownian   84
    fractionary Brownian   4, 84
multifractals   3, 59-71, 124-125, 188-192
network   210-213
    hydrographic   195-197
noise (Brownian)   85
noise (white)   42
Olbers (paradox)   160
orbit (stable)   187
oscillator(s)   31-33
    harmonic   31, 37
    coupled harmonic   33, 37
palaeo-intensity   144-145
palaeontology   232-233
pendulum
    simple   32, 33, 37
    magnetic   34-37
percolation   103-106, 210-219
phases' space   3, 31-33, 35-45, 48, 51-56, 59, 60, 65, 80, 182, 224
Poincaré   1-3, 17
Poincaré section   45, 53-56, 58, 71, 72, 75,-78, 80, 123, 157, 184, 186
Poisson law   107-113, 116
power law   7, 86, 107-113, 116, 233

predictor   152-153
probability   6, 91-95, 203-209, 225
problem
    direct   149, 150
    inverse   149, 150
product of fractal sets   28, 29
projection of fractal sets   27, 188, 189
    projection in $\mathcal{R}^2$ 27, 28
    projection in $\mathcal{R}^n$ 28
quaternions   177
randomisation   21
renormalisation   6, 90, 210
renormalisation group   6, 90
Richardson (measure)   195
Ruelle-Takens theory   76, 77
sand heap   126-129, 171, 209
scale   17, 21, 46, 57, 58, 66-68, 86, 94, 95, 99, 101, 107, 109, 116, 128
scale (Cox)   138, 140-143
scale invariants   109, 128, 129
section (Poincaré)   45, 53-56, 58, 71, 72, 75,-78, 80, 123, 157, 184, 186
seismicity   107, 117
seismology   81, 219-223
self-affine (fractal)   83
self-organised criticality (SOC)   4, 126-129, 154-156, 171-172
sensitivity to the initial conditions (SIC)   35, 40, 183
Sierpinski
    Sierpinski's carpet   25
    Sierpinski's square   164
    Sierpinski's triangle   24, 25, 163
Sinaï (billiards)   169
singularity index   62, 65
skeleton   105
speciation   232
spectrum

      Fourier spectrum   39
      singularity spectrum   61, 67, 124, 125
spreading   82, 193
system
      discrete system   53
      dynamic system   6, 31-37, 169-171, 181-183
tectonics   101-103, 113-116, 203-219
threshold
      instability   128
      percolation   103, 210-219
time series   151-154
topology   199
toroid
      T^2 toroid   35, 75, 76
      T^3 toroid   76, 77
trajectory (phase)   31, 32, 35, 36, 38, 39, 41, 42, 52, 75, 182
transform
      affine   83, 160, 161
      baker's   169, 170
      Fourier   68, 83
      Legendre   63, 192
      wavelet   3, 59-71
translation   161
tremors   124
union   10, 11
variance   84
volcanic eruptions   118-125
volcanic tremors   124
Von Koch   23-24
Walker-Lewis (law)   114
wavelet   59-71
      Mexican hat   69
      Morlet   68-69
      wavelets constant by parts   69, 70